Resource and Environmental Sciences Series

General Editors:
Sir Alan Cottrell, FRS
Professor T.R.E. Southwood, FRS

Already published:
Environmental Economics — Sir Alan Cottrell
Energy Resources — J.T. McMullan, R. Morgan and R.B. Murray
Food, Energy and Society — David and Marcia Pimentel
Environmental Biology — E.J.W. Barrington
Environmental Chemistry — R.W. Raiswell, P. Brimblecombe, D.L. Dent
and P.S. Liss
Environmental Toxicology — John H. Duffus
Minerals from the Marine Environment — Sir Peter Kent

Ecology for Environmental Sciences:
Biosphere, Ecosystems and Man

J. M. Anderson

Department of Biological Sciences,
University of Exeter

A HALSTED PRESS BOOK

JOHN WILEY & SONS
New York

© J.M. Anderson 1981

First published 1981 by
Edward Arnold (Publishers) Ltd, London
Published in the U.S.A. by Halsted Press, a Division of
John Wiley & Sons, Inc., New York.

Library of Congress Cataloging in Publication Data

Anderson, J.M.
    Ecology for environmental sciences.

    (Resource and environmental science series)
    "A Halsted Press Book."
    Includes bibliographical references and index.
    1. Ecology.   I. Title.   II. Series.
OH541.A495   1981   304.2   81-3349
ISBN 0-470-27216-3   AACR2

Printed in Great Britain

Preface

The presentation of another general ecology book requires some justification in view of the number of established texts available. In recent years our knowledge of ecology has developed rapidly on so many fronts that there are practical restraints on how much new material can be incorporated in a general textbook while still retaining essential background material. For example, the three editions of one environmental science text so exactly conform to an exponential growth curve of weight or page numbers that it is possible to predict the publication of a massive 55 kg tome by the mid 1990s if its popularity continues! Recently, however, there has been a trend towards smaller, specialist texts which can be assembled by students to provide a more comprehensive coverage of ecology in its various aspects than is feasible for one or two authors to present.

This book therefore, is not intended as a comprehensive introduction to ecology but as a core text for first or second year courses illustrating different approaches to ecological topics which can be expanded using the literature cited.

In the first half of the book the structure and functioning of ecological systems is described from the viewpoints of the various contributory disciplines of the environmental sciences: the biosphere and ecosystems may be considered by geologists/geographers with a minimum of biological detail, while the study of communities and populations is more the realm of the biologist. At each scale of organization a holistic approach is adopted and then the component processes are considered; in each case using practical examples and avoiding unnecessary theoretical detail. The effects of man on his environment are emphasized throughout.

In the second part, the processes of primary production, secondary production and decomposition are examined in greater detail including crop production, livestock and alternative protein sources and applied aspects of decomposition. Much of this material is not found in general ecology texts and I have found that it stimulates considerable interest and discussion among students.

Exeter
1980

J. M. Anderson

Preface

Contents

1 Introduction: The Organization of Ecological Systems

Ecology is essentially the study of the relationship between organisms and their *environment*. The term environment is often used to simply refer to the physicochemical parameters of the *habitat* (place) where the organism lives. Species, however, do not exist in isolation from one another and therefore their natural environment includes other organisms. These biotic dimensions of the environment define the ecological components of the various contributory disciplines such as botany, zoology, genetics, geography and biochemistry.

An organism and the environment to which it is adapted is an inseparable ecological unit. The ultimate goal in ecological science, the understanding of why living systems exist in their observed form and functional relationships, can therefore only be answered by a holistic approach (the functional relationships between parts and the whole system). Often it is sufficient to interpret particular ecological phenomena by proximate factors, for example the distribution of marine organisms in relation to salinity or tidal levels. It is necessary, however, to recognize the existence of competition and predation as the ultimate biotic factors which have moulded this pattern over an evolutionary time scale. Without this appreciation it is likely that false assumptions or predictions will be made regarding the ecological consequences of natural or man-induced changes in the environment.

There have been some one million plant species and 1½ million animal species described in the world to date. Only a small fraction of these have been subject to detailed ecological investigation and our knowledge of how these species interact decreases with increasing numbers of species. Few studies have been made of more than three interacting species. The largely unknown effects of multispecies interactions makes it essential in the study of ecology, at the present time, that the whole is regarded as greater than the sum of the parts.

It might be questioned, on this basis, as to how a holistic approach can be adopted to such disparate problems as the local effects of a pesticide on non-target species and the ecological consequences of car exhaust emissions affecting the stratosphere. The answer is that it is a problem of defining the question at an appropriate scale of organization of ecological systems, ranging from the planet, biosphere, ecosystem and community to the local population. If we view these ecological systems as a series of 'black boxes' nested inside one another then the functional properties of a box can be measured as inputs and outputs, even if the internal organization is unknown. If the problem is not resolved at that particular scale, then the box can be 'opened' to reveal emergent properties of interacting subsystems. Alternatively, the 'box' may be recognizable as an interactive part of a larger system. We will see the application of this concept to the investigation of ecosystem structure and functioning (Chap. 3).

The largest system with which we are concerned is the *planet Earth*. The resources of the earth are finite; significant amounts of material are neither lost nor gained across the atmospheric boundary. Radiant energy from the sun, however, enters the system and is re-radiated to space as heat in a one-way flux. The earth is essentially a closed system for materials and little information could be gained on the internal organization of the planet by studying element fluxes across the boundaries. Some information on the physical nature of the earth's core could be deduced from the slightly greater heat output than solar energy input.

Within the boundaries of the earth's atmosphere we can recognize several functional compartments. The highest order of ecological system is the *biosphere* − the region of the planet in which organisms live and reproduce. The physicochemical environment of the biosphere is the solid earth (*lithosphere*), water bodies (*hydrosphere*) and the *atmosphere*. Many naturally occurring elements pass from the abiotic environment to living organisms, for growth and maintenance, and are finally returned to the mineral pool by excretion and decay. These cyclical fluxes of elements, the *nutrient cycles*, are integrated into much larger and longer term physical cycles of elements between the lithosphere, hydrosphere and atmosphere referred to collectively as *biogeochemical cycles* (see p. 15). Energy is not cycled but is transduced from one form to another by biological systems according to the laws of thermodynamics. Sunlight energy is 'fixed' by the photosynthetic activity of plants (autotrophs) and released as heat by the metabolic activity of the plants themselves as well as heterotrophs (animals, fungi and most bacteria).

Living material is not evenly distributed in the biosphere but is aggregated both qualitative and quantitatively into different groups of animals and plants according to the climate and other factors. The gross assemblages of these units are the *biomes* − Tundra, Tiaga (Northern Coniferous Forest), Temperate Deciduous Forest, Grasslands, Tropical Rain Forest, and so forth. The term is principally used to refer to terrestrial systems but major divisions of freshwater bodies and the oceans are also recognizable.

The biomes are not homogenous entities. Within the deciduous forest biome, for example, one can recognize a number of different types of woodland formed by a predominance of one or more species. In broad functional terms the beech, oak, hemlock or tulip poplar woodlands are rather similar to one another but differ from grasslands. Similarly, tall or short grass prairies and different types of savannah show closer functional affinities to one another than woodlands. Thus within a biome we can define a typical assemblage of plant and animal types or a series of specific assemblages of animal and plant species. The unit of plants and animals which is functionally integrated through nutrient and energy fluxes is the *ecosystem*. The environment of an ecosystem is formed by adjacent ecosystems with which it interacts through fluxes of materials across their boundaries. As we will see (p. 36) the measurement of these fluxes can provide considerable insight into the internal dynamics of the ecosystem and also allow a more precise definition of its identity.

Within the ecosystem boundary there are three functional compartments (Fig. 1.1) — the green plants (*autotroph subsystem*), the animals, together with their predators, which feed on living plants (the *herbivore subsystem*) and the organisms decomposing dead plant and animal remains, together with their predators (the *decomposer subsystem*). Nutrient elements taken up from the soil and air by plants are synthesized into tissues using light energy (*primary production*). The total amount of material and energy fixed by the autotrophs is called *Gross Primary Production* (GPP). Some proportion of the GPP is respired (R) by the plants and the remaining *Net Primary Production* (NPP = GPP − R)

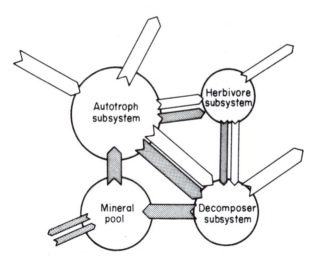

Fig. 1.1 Major components of ecosystem structure and functioning. Nutrients (tinted arrows) circulate between the autotroph subsystem (green plants), the herbivore subsystem (herbivores and their associated predators), the decomposer subsystem (animals and microorganisms) and the mineral pool. Energy (open arrows) enters the ecosystem as solar radiation and leaves as metabolic heat in a one-way flux. There is little energy exchange between adjacent ecosystems. Larger nutrient than energy fluxes may occur across ecosystem boundaries but these are always small in comparison to the internal cycles.

is available to the heterotrophs. The production of tissues by the growth and reproduction of heterotrophic organisms in the herbivore and decomposer subsystems is known as *secondary production*. Ultimately all material is processed by the decomposers and the elements released in a mineral form suitable for re-utilization by plants. It is useful from an ecological point of view to distinguish between relatively unavailable nutrients in a *storage pool* and the active, exchangeable reservoir or *exchange pool*. The capital of nutrient elements accumulated in living and dead tissues (*standing crop*, as distinct from *biomass* which consists of living material only*) is referred to as the *utilized pool*. An essential feature of an ecosystem is that while nutrients are recycled within its boundaries, energy enters as light and leaves as metabolic heat in a one-way flux.

The total species assemblage of the three subsystems as well as that of the individual subsystems is termed a *community*. This rather arbitrary term may also be used for smaller groups of interacting species. The extent to which a specific community, or communities, is associated with an ecosystem depends upon the coherence of its boundaries. In a continuous gradient extending from one ecosystem type to another it may be very difficult to delimit discrete plant or animal communities. Nonetheless, we can usually functionally define a community and measure its gross activity (e.g. decomposition) as well as its interaction with other communities (e.g. the interactions between plant and animal communities).

At a finer level of ecological resolution, we see that a community is composed of *populations* of individual species. Once we have physically defined the limits of the population we can deduce something about its dynamics or resistance to disturbance by measuring inputs and outputs of energy or nutrients (e.g. for an ant nest) or immigration and emigration. The internal dynamics of a population are primarily determined by balance of birth rates and death rates. The ultimate factors regulating these parameters are other organisms (competitors, predators and parasites) as well as the abiotic environment, particularly the weather (short term temperature, wind and rain conditions) and climate (long term averages of the same measurements).

Most interactions between organisms and their environment, both biotic and abiotic, tend to reach a state of equilibrium over ecological or evolutionary time scales. Physicochemical processes in the biogeochemical cycles, such as the composition of the earth's atmosphere, also establish equilibrium points under natural conditions. The mechanism by which these phenomena are regulated is known as *homeostasis* and operates through negative and positive feedback loops controlling deviations from an equilibrium point. A simple example of homeostasis operating in an

---

*A number of other uses of these terms will be found in the ecological literature. It is convenient, however, to use them to distinguish between living and dead material, for example grass in a tussock. The term biomass literally means 'quantity of life'.

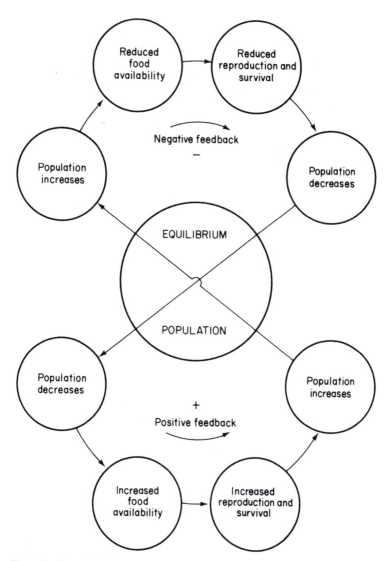

Fig. 1.2 Hypothetical model of the homeostatic processes regulating an animal population through food availability. The extent to which a real population remains constant in numbers or fluctuates about an equilibrium value is determined by the action of both intrinsic factors (life history, age structure and intraspecific competition for resources) and extrinsic factors (weather, predators, disease and interspecific competition), many of which directly or indirectly affect the quantity and quality of the food supply.

animal population is shown in Fig. 1.2. A population with a predominance of positive feedback would show exponential growth, and then die out as resources become limiting. The operation of strongly negative feedback loops would cause the population to decline to extinction. Perfect density dependent homeostasis, in which positive and negative feedback loops respond instantly to changes in state would result in stable equilibrium, as with a sensitive thermostat. Ecological systems rarely, if ever, achieve stable equilibrium and exhibit varying degrees of fluctuation about an environmentally determined equilibrium point with respect to time. This is an inherent consequence of the complexity of the regulatory mechanism in the environment and the time delays in their operation.

Homeostatic mechanisms operate at all levels in ecological systems, from the population to the biosphere, and their identification is of paramount inportance in understanding how an ecological system will respond to disturbance. The extent to which an ecological system will withstand different degrees of perturbation may be described in terms of *neighbourhood (or local)* and *global stability* (Lewontin, 1969). These concepts are represented in Fig. 1.3. A system returning to equilibrium after small disturbances exhibits local stability, for example grass biomass in a mown meadow. If large perturbations are accommodated by the system's homeostasis, it is said to have global stability. Large and repeated additions of fertilizers to Crecy Lake (Canada), for example, caused only temporary effects and induced *eutrophication* (nutrient enrichment) proved reversible over an 18 year period (Smith, 1969). The same phenomenon, on an even larger scale, has been observed for Lake Washington which recovered from a highly polluted condition following the cessation of sewage disposal in the lake in 1967 (Edmondson, 1969). Both of these lake systems evidently have high global stability (c.f. Lake Erie, p. 55).

Communities may exhibit local but not global stability. Animals and plants tend to alter their environment so that it becomes more suitable for other species. For example, the formation of soil by mosses colonizing bare rocks allows grasses and other plants to become established, eventually eliminating the moss, and finally trees will follow grasses. This is the process of *ecological succession* and the final community, the climax, will show global stability: if a woodland is cut down or burnt and reverted to grassland it is likely to be re-established under the same environmental conditions.

In the following chapters we investigate these different scales of ecological organization in more detail.

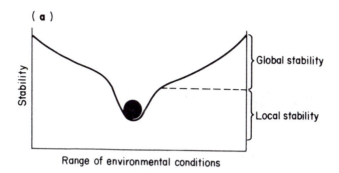

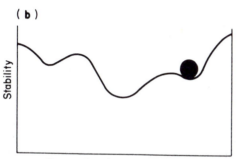

Fig. 1.3 Local and global stability concepts. The ecological system (population, community or ecosystem) is shown as a sphere located on a topographic surface representing a range of environmental conditions. The depressions in the surface represent different levels of stability imparted by homeostatic processes. In situation **(a)** the system will return to the same configuration following small (local) or large (global) perturbations. Situation **(b)** is locally stable but perturbations beyond a critical level will cause a displacement to a different configuration. Ecological succession is directional and could therefore be represented as a sloped topographic surface with regions of local stability. The climax would be at the lowest point and exhibit local and global stability.

2   The Biosphere

Biological activity is dependent upon the supply of energy, water and nutrient elements. The distribution and abundance of these requirements, at the scale of the biosphere, is governed by world patterns of radiation input from the sun, the hydrological cycle and the geological cycles of minerals in the earth's crust. These abiotic processes will be described briefly before we consider the interaction between the biosphere and its physical environment.

Energy, Climate and Weather

The Earth spins on an axis of rotation which is tilted 23.5° from the perpendicular through its elliptical orbit round the sun. The net effect of this inclination is to produce strong gradients in the incident solar radiation from the poles to the equator which vary seasonally with the orbital periodicity of the Earth (Fig. 2.1). In late June (mid-summer solstice) the Northern Hemisphere is tilted towards the sun and the entire area north of the Arctic Circle receives 24 hours of illumination. The intensity of illumination is lower at the poles than the equator because of the shallow incident angle of radiation from the sun and the depth of atmosphere which the light has to penetrate. The Southern Hemisphere has the summer solstice in December when the Northern Hemisphere is in mid-winter and the North Pole experiences 24 hours of darkness. As the Earth moves round its orbit, the seasons vary between these extremes so that on the equinoxes (in March and September) there is continuous twilight at the poles, and day and night are approximately equal in temperate and tropical regions.

The amount of solar energy reaching any point on the outer limits of the atmosphere is therefore a function of latitude. Once it enters the atmosphere the energy is dispersed in a number of ways which depend on cloud cover, dust content of the atmosphere and a number of similar diurnal and seasonal physical variables. A generalized scheme for the fate of incoming solar radiation is shown in Fig. 2.2. Averaged over the year and around the globe about 25–30 per cent of the sun's energy is reflected back into space by clouds and the gases of the atmosphere. Clouds, dust and gases absorb a further 25 per cent of the radiant energy, which reaches

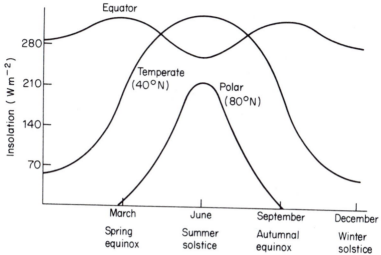

Fig. 2.1 Annual variation of insolation in equatorial, temperate and polar regions of the Northern Hemisphere. (After Rumney, G. R. (1970). *The Geosystem: Dynamic Integration of Land, Sea and Air*. William Brown, Dubuque, Iowa.)

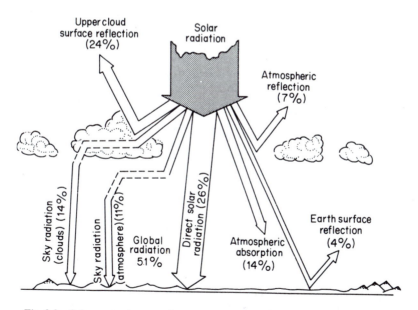

Fig. 2.2 Solar energy input to the earth and atmosphere. (From Rumney, G. R. (1970). *The Geosystem: Dynamic Integration of Land, Sea and Air*. William Brown, Dubuque, Iowa.)

the ground in diffuse form, while a similar proportion is directly radiated to the earth's surface. A small proportion of the radiation is reflected back into space, the amount depending upon the reflectivity or albedo of the surface. Snow, for example, may reflect back 80 per cent of the solar energy received, and consequently warm slowly, while grassy surfaces may reflect 20 per cent and dark soil as little as 10 per cent of the radiation.

Much of the energy absorbed by land and water bodies is dissipated by the evaporation of water. When this moisture condenses, the latent heat warms the atmosphere. Further heating of the atmosphere is contributed by the last 20 to 25 per cent or so of the energy not accounted for in our budget.

To understand the energy transformations which take place in the atmosphere it is necessary to consider the distribution of wavelengths in the radiation reaching the earth. Approximately 41 per cent of the energy is in the visible light part of the spectrum (400–710 nm), 50 per cent is in the infrared (710–3000 nm) and 9 per cent in the ultraviolet. Much of the visible light reaches the ground, together with about 30 per cent of the infrared and most of the near ultraviolet radiation (wavelengths just shorter than visible light). Some 20 per cent of the infrared is absorbed by dust, carbon dioxide and water vapour in the atmosphere below 10 km (Mount Everest is 8.9 km high) and transformed into heat. We are finally left with about 1–3 per cent of the ultraviolet between 0.18 and 0.30 μm. Radiation of these wavelengths damages living organisms but, fortunately for present life on earth, it is absorbed by ozone (O_3) in the atmosphere above 30 km. The surface of the prebiotic earth was exposed to intense short wave ultraviolet radiation as insignificant oxygen was present for the ozone layer to form. The development of present oxygen levels in the atmosphere through the photosynthetic activity of plants represents one of the earliest, and perhaps the most significant, ways in which the biosphere has modified its physical environment. Only a tiny proportion of the total incoming radiation has been used by plants to effect this transformation. The overall picture, therefore, is of an atmosphere relatively impenetrable by short wave radiation but gaining energy from long wave radiation, particularly the re-radiation of heat absorbed by the earth's surface. This results in the maintenance of fairly high temperatures near the ground, a phenomenon known as the 'greenhouse effect'.

Water and the Hydrologic Cycle

The principal flows in the hydrologic cycle involve the evaporation of water from land and sea by solar radiation, the transport of atmospheric water from one place to another, precipitation as rain, hail, sleet or snow, and runoff from the continents to the sea. At the present time the mean oceanic evaporation rate of 1200 mm m^{-2} yr^{-1} exceeds precipitation by about 100 mm^{-2} yr^{-1} while on land precipitation of 710 mm^{-2} yr^{-1}

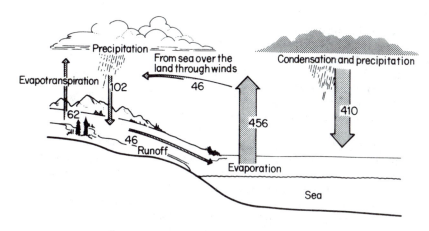

Fig. 2.3 The hydrologic cycle (flux rates in 10^{12} t yr^{-1}). The cycle of water through the biosphere requires worldwide evapotranspiration and precipitation to be equal. Almost 95 per cent of the water on earth is chemically bound in rocks and does not cycle. A small amount of this reserve is released as juvenile water by volcanic activity and compensates for water bound into the reserve pool and losses of hydrogen to space. (Data from Budyko, M. I. (1974). *Climate and Life*. Academic Press, New York.)

exceeds evaporation by nearly 240 m^{-2} yr^{-1}. Thus there is a net cyclical movement of water from the sea to the land as vapour and from the land to the sea as river discharge (Fig. 2.3). The atmosphere contains only 0.7 per cent of the free water on earth with 99.3 per cent located in the oceans and ice caps. The turnover time in the atmosphere is correspondingly small: in the order of 9 days compared with several weeks to a year in the soil surface and 120–3000 years in the oceans. The balance between precipitation and evaporation varies widely from one continent to another (Table 2.1) and is a measure of the water available for agriculture, domestic and agricultural uses. South America has the largest runoff of any of the continents and the discharge from the Amazon River, which drains the wettest third of the land surface, amounts to about a seventh of the global discharge.

Table 2.1 Average water balance of the continents. (After Budyko, 1974.)

	Precipitation (mm yr^{-1})	Evaporation (mm yr^{-1})	Runoff	
			(mm yr^{-1})	(10^9 t yr^{-1})
Africa	690	430	260	7 700
Asia (including USSR)	600	310	290	13 000
Australia	470	420	50	380
Europe	640	390	250	2 200
North America	660	320	340	8 100
South America	1630	700	930	16 600

The hydrological cycle forms, on a meaningful time scale for man, a natural distillation process which is of great significance for the cycling of many mineral elements. It is estimated that 2.73×10^9 t yr^{-1} (t = tonne) of dissolved materials are brought into the oceans from the land mass by rivers and a further 9.3×10^9 t yr^{-1} of suspended solids before extensive disturbance of the land surface by man (Whittaker, 1975). Current estimates of particle erosion rates are about 24×10^9 t yr^{-1}.

The global distribution of precipitation results from complex patterns of energy fluxes in the atmosphere, lithosphere and hydrosphere. The orientation of the earth's surface to the sun produces large scale atmospheric movements due to the differential heating and cooling of different parts of the globe. In the tropics, where the earth's surface is perpendicular to the rays of the sun for most of the year, the heating of the ground forms convection currents in the atmosphere. These updrafts of air cool as they rise and the moisture condenses out, forming the characteristically overcast conditions of the humid tropics. At the poles the air cools, descends in altitude and then moves across the earth's surface towards the tropics. The Coriolis forces, caused by the rotation of the earth, deflect these surface winds to the right in the Northern Hemisphere and to the left in the Southern Hemisphere, forming the Trade Winds.

The large scale circulation of air contains a number of smaller cells, the most important of which descend at about latitude 30° and flow northward and southward. The descending air is dry and these latitudes (North and South Africa, Australia and South America) have low rainfall, clear skies and high temperatures away from coastal regions. The warm air moving towards the poles picks up moisture vapour from the land and sea. This precipitates when the warm air meets the cold air flow from the poles and forms a rain belt extending across the temperate regions.

This basic climatic pattern is modified seasonally, as the cells are shifted northward and southward by the relative positions of the earth and sun, and topographically by mountain ranges. As air rises to pass over mountains, it cools and the moisture it contains is precipitated as rain or snow. On the lee side of the mountains the dry air descends and forms an arid 'rain shadow'.

The somewhat simplistic pattern of global climate given above is far removed from the complexity of factors governing actual conditions. A large number of feedback loops regulate climate and effectively stabilize it against small changes in the energy fluxes within the system or in radiation input from the sun (Fig. 2.4). The high thermal capacity of the oceans, for example, buffers seasonal changes in temperature on the land masses. As we will see, however, there is evidence that modern man may have disturbed the long-established radiation balance of the earth through modification of the atmospheric gas composition and an enhancement of the 'greenhouse effect'.

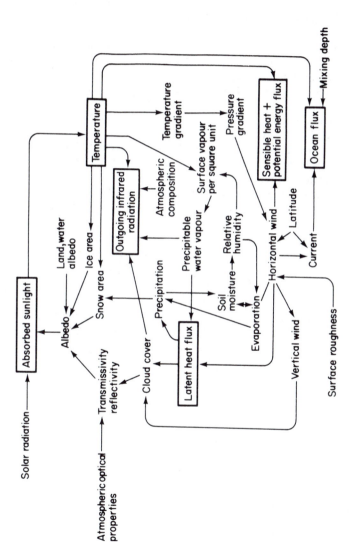

Fig. 2.4 Some of the feedback loops governing global climate. (After Kellog, W. W. and Schneider, J. H. (1974). *Science*, **186**, 1163–72. Copyright 1974 by the American Association for the Advancement of Science.)

Geologic Cycles

The principal processes in the geologic cycle of materials are shown in Fig. 2.5. Rock which is exposed on the surface of the earth's crust is weathered away by physical and chemical processes. The particles and soluble minerals are predominantly moved downhill under the influence of gravity and over long periods of time materials eroded from high elevations are deposited in lakes, rivers, the seas and low-lying land as sediments. As the layers of sediment become more deeply buried, high temperature and pressures cause physical and chemical changes which result in the formation of sedimentary rocks. Carbonates, from shells or chemically precipitated carbon dioxide, which are incorporated into sediments, form limestones. Under some extreme geological conditions, such as the uplifting of land masses, extremes of heat and pressure are generated which result in further changes and the formation of metamorphic rocks, including slate and marble. Igneous rocks are the most abundant rocks in the earth's crust and are formed by the melting of sedimentary and metamorphic rocks. The repeated melting and solidification of magma has resulted in a stratification of the crust into the surface granitic rocks of

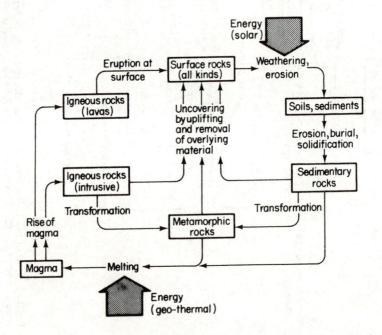

Fig. 2.5 A schematic representation of the major processes in the geologic cycle. See text for details. (After Ehrlich, P. R., Ehrlich, A. H. and Holdren, J. P. *Ecoscience: Population, Resources, Environment*. W. H. Freeman and Company, Copyright © 1977.)

the continental masses and the heavy basaltic rocks, containing more iron compounds, of the ocean floor. The uplifting of the land masses to form mountains or the eruption of volcanoes completes the sedimentary cycles of minerals over periods of tens to hundreds of millions of years. Over an ecological time scale minerals deposited on the deep sea sediments are effectively withdrawn from circulation.

The energy which drives the geologic cycles has two distinct origins. The geophysical processes (the production of magma, volcanic activity and the uplifting of land masses) are driven by heat produced by the decay of potassium, uranium and thorium isotopes in the core of the earth. Those parts of the cycles which take place on the earth's surface, such as erosion, weathering and the transport of sediments, are powered by solar energy manifested in the forms of heat extremes, wind and hydrological activity.

Biogeochemical Cycles

The biosphere is composed of some 1.8×10^{12} tonnes of living tissue. This seemingly vast amount of material represents only a thin layer over the surface of the globe, proportionately thinner than the skin of an apple. If all the material was evenly distributed it would be less than 1 cm thick and weigh about 3.6 kg m^{-2}. This average hides considerable variation in the distribution of biomass ranging from about 45 kg m^{-2} in tropical forests to 0.003 kg m^{-2} in the oceans (see Table 5.2). The arid deserts, polar ice caps and deep caves are more sparsely colonized, sometimes only by microorganisms. Wherever organisms live in the biosphere, however, they have certain requirements which are qualitatively similar for all forms of life. These requirements are the nutrient elements for tissue synthesis and the energy for metabolic maintenance.

The main groups of elements required by living organisms, and some of their functions, are shown in Table 2.2. Both the abundance of these elements in the atmosphere, lithosphere and hydrosphere, and their availability are critical determinants for the structure and functioning of living systems.

The first group of elements, carbon, hydrogen and oxygen, form most of the material in the biosphere both as water (about 75 per cent of the biomass) and as dry matter. The dry matter is predominantly CH_2O polymers in plant materials, principally cellulose. Animal tissues have lower carbon contents. The formation of carbohydrates through photosynthesis and their breakdown by respiration closely integrates the biosphere with the mineral pools of these elements.

The macronutrients include nitrogen which is the fourth most abundant element in the biosphere at about 2 atoms per cent. The atmosphere forms a vast reservoir of nitrogen (about 75 per cent by volume) but it is a very inert element. Animals and the majority of plants are ultimately dependent upon nitrogen fixing microorganisms for a supply of nitrogen in a similar

Table 2.2   Elements necessary for life, and some of their functions.

Category	Element	Symbol	Some known functions
Major constituents (20–60 atoms per cent)	Hydrogen	H	Universally required for organic compounds of cell
	Carbon	C	
	Oxygen	O	
Macronutrients (0.02–2 atoms per cent)	Nitrogen	N	Essential constituents of proteins and amino acids
	Sodium	Na	Important counter-ion involved in nerve action potentials
	Magnesium	Mg	Cofactor of many enzymes, e.g. Chlorophyll
	Phosphorus	P	Universally involved in energy transfer reactions and in nucleic acids
	Sulphur	S	Found in proteins and other important substances
	Chlorine	Cl	One of major anions
	Potassium	K	Important counter-ion involved in nerve conduction, muscle contraction, etc.
	Calcium	Ca	Cofactor in enzymes, important constituent of membranes and regulator of membrane activity.
Micronutrients (trace elements) (less than 0.001 atoms per cent)	Boron	B	Important in plants, probably as cofactor of enzymes
	Silicon	Si	Found abundantly in many lower forms such as diatoms
	Vanadium	V	Found in respiratory pigments of some lower animals
	Manganese	Mn	Cofactor of many enzymes
	Iron	Fe	Cofactor of many oxidative enzymes, e.g. haemoglobin
	Cobalt	Co	Constituent of vitamin B_{12}, required for N fixation
	Copper	Cu	Cofactor of many oxidative enzymes
	Zinc	Zn	Cofactor of many enzymes, e.g. insulin
	Molybdenum	Mo	Cofactor of a few enzymes, particularly nitrogenase (N fixation)
	Iodine	I	Constituent of thyroid hormones which regulate growth and development in vertebrates

manner to the dependence of heterotrophs upon autotrophs as a source of available energy. The remaining macronutrients, with the partial exception of sulphur, are involved in sedimentary cycles so that the age and composition of the parent rock influences their abundance and availability in soil. Phosphorus, for example, is not only scarce in the lithosphere (less than 0.1 atoms percent) but a major proportion of the phosphorus in most soils is unavailable to plants.

The final group, the micronutrients or trace elements, are required in very small quantities but their absence or deficiency can have deleterious effects on animals and plants. Many organisms have highly specific trace element requirements (see Table 2.1). Molybdenum deficiencies, for example, limit nitrogen fixation by free-living bacteria in some soils of Australia and the U.S.A. Many of the trace elements are important industrial minerals and their release by man into the environment in super optimal concentrations can have harmful effects. Iodine is one of the heaviest elements (atomic weight 127) required by living organisms but is only in the 53rd position on the Periodic Table. Heavier elements such as mercury and lead are biologically active to varying degrees but do not have a metabolic role and are usually inhibitory to biological processes (p. 93).

All the elements given in Table 2.2 have characteristic biogeochemical pathways but it is only possible to consider here the gaseous cycles of carbon, hydrogen, oxygen and nitrogen, the sedimentary cycle of phosphorus and the pathways of lead in the environment affected by man's industrial activities.

Gaseous Cycles

Carbon, hydrogen and oxygen
The cycles of carbon, hydrogen and oxygen are so intimately bound together (Fig. 2.6) that it is appropriate to consider them together. Detailed considerations of the individual cycles are available in current reviews such as *Biosphere* (1971).

The object of the following section is to emphasize the interaction and balance of the cycles under natural conditions and the ways in which man's activities have been accommodated by, or have exceeded, the natural homeostatic processes.

The fluxes of carbon, hydrogen and oxygen represent the major link between the biosphere and its abiotic environment through the complementary processes of photosynthesis and respiration:

$$6\,CO_2 + 6\,H_2O + \text{energy} \underset{\text{respiration}}{\overset{\text{photosynthesis}}{\rightleftharpoons}} 6\,C_6H_{12}O_6 + 6\,O_2$$

There is considerable variation in the proportion of gross primary production utilized as plant respiration, ranging from 20 per cent to 75 per cent

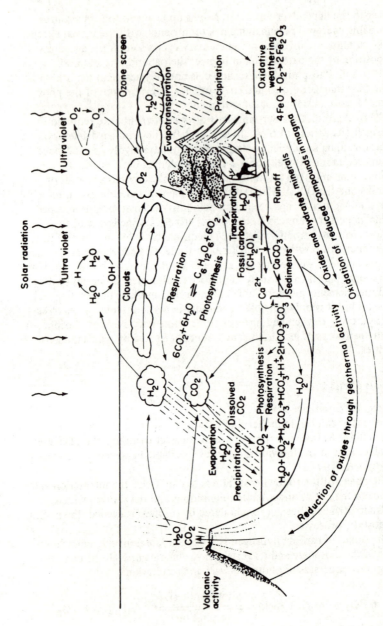

Fig. 2.6 Interrelationships between the carbon, hydrogen and oxygen cycles. All three elements occur in many different forms and so only the major pathways are shown.

depending upon plant type and region, but a global average is in the order of 50 per cent. Thus 50 per cent of the CO_2 and H_2O fixed by plants is respired, using about 50 per cent of the oxygen they produce. The remaining 50 per cent of GPP, net primary production, is respired by the heterotrophs of the herbivore and decomposer subsystems. This balance between producers and consumers, which is implicit in our understanding of the biosphere today, has not always existed but evolved over millions of years.

The earth was formed some 4.5 thousand million years ago by the coalescence of material from the solar nebula. The solid phase consisted principally of a mixture of silicates together with iron and sulphides. These materials differentiated according to their masses into the metallic core of the earth, a mantle of dense oxides and a thin crust of silicious rocks. Water was not present in a free form but as hydrated minerals such as mica in the earth's crust. The primary atmosphere was strongly reducing and mainly consisted of hydrogen. The relative paucity of hydrogen and the gases helium, neon, argon and xenon in the atmosphere today has led to speculation that the primary atmosphere was stripped away in pregeological history, possibly during the formation of the Moon. The primordial atmosphere and oceans were of secondary origin resulting from volcanic activity. Estimates have been made that, at present levels of activity, approximately 2.6 cubic kilometres of solids are added to the land mass each year. Over 3 to 4 billion years this would be sufficient to account for the present continental land mass. Water vapour forms about 97 per cent of volcanic gases by volume and, over the millenia, water released from rocks has condensed to form a major proportion of the hydrosphere. Nitrogen, CO_2, NH_3, SO_2 and traces of other gases were also released from the lithosphere by geothermal activity. Traces of oxygen were present in the atmosphere, formed by photodissociation of water, but at insufficient levels to result in the formation of an ozone layer. The land mass and surface waters were consequently exposed to intense ultraviolet radiation.

The first life forms of 3000 million years ago were simple unicellular organisms (prokaryotes) which evolved at the bottom of shallow pools or seas of 10 m or more, where they were protected from ultraviolet radiation. These early forms of life were probably chemoheterotrophs i.e. they depended upon simple, abiotically synthesized organic compounds as energy and nutrient sources. Chemoautotrophic bacteria also evolved which were able to derive energy for biosynthesis from the oxidation of reduced inorganic compounds, such as hydrogen sulphide (H_2S) to sulphur. The world's great elemental sulphur deposits attest to the success of this group but today their significance in the biosphere is negligible. One possible reason for this was the appearance of the photoautotrophs which use light as an energy source and which, judging from fossil evidence, resembled nitrogen fixing blue-green algae. The fixation of nitrogen probably accelerated the accumulation of photosynthetic biomass and hence the rate of oxygen release into the atmosphere. Massive deposits of iron oxides dating from more than 2.7 billion years ago mark the appearance of significant levels of free oxygen.

Over the next 1 to 1.5 billion years oxygen accumulated in the atmosphere to about 1 per cent of present day levels and over this period it is believed that the ozone screen became effective and oxidative metabolism evolved, though the precise timing of these events is contentious (Cloud and Gibor, 1970). Oxidative metabolism releases 30 to 50 times more energy per substrate molecule than anaerobic (fermentative) respiration and results in a greater excess of gross primary production over autotroph respiration. The increased net primary production available to heterotrophs, the presence of abundant nitrogen fixing microorganisms and the protective ozone layer are likely to have provided a positive feedback on the development of the biosphere. Both plants and animals evolved rapidly during the Cambrian and by the late Silurian and Devonian the seas, judging by the fossil record, were swarming with life. During the Carboniferous Era (340–280 million years ago) the interior of the continents was probably devoid of vegetation but coastal regions were dominated by vast swamps of giant ferns and horsetails. The remains of these primeval swamp forests formed a major part of our present day fossil fuel reserves. Organic materials were prevented from decomposing by the anaerobic conditions in the swamps and deeper water sediments. As they became silted over and compressed, they were transformed into oils, tars and natural gas deposits as well as coal (see Raiswell *et al.*, 1980 for details). Organic matter still accumulates in swamps, bogs and river deltas but on an insignificant scale compared with the Palaeozoic Era.

The withdrawal of vast amounts of reduced carbon from the biological cycles accelerated the accumulation of oxygen in the atmosphere. The total amount of carbon in these deposits is equivalent to the production of some 13×10^{15} tonnes of oxygen by photosynthesis (and photodissociation) during the history of the earth. The atmosphere contains approximately 0.8×10^{15} tonnes O_2 and 0.2×10^{15} tonnes O_2 is dissolved in the hydrosphere. Thus only about 7 per cent of the O_2 produced exists in a free state today and the remainder has accumulated in reserve pools of mineral oxides, sulphates and carbonates in the lithosphere and as H_2O in the hydrosphere. The turnover of carbon and oxygen in the carbonate pool takes place over a geological time scale through weathering and volcanic activity but the mobilization of fossil carbon has been more rapid in modern times. It is estimated that the burning of oil and high grade coal deposits results in the release of reduced carbon at about 2000 times the fossil carbon storage rate. This oxidation requires the removal of two atoms of oxygen from the atmosphere for each reduced carbon atom. This seemingly high rate of oxygen consumption has not, in fact, resulted in significant changes in the oxygen content of the atmosphere over the last 100 years or so. Calculations of total fossil fuel consumption over the period 1910 to 1970 suggests that, at most, a decrease of 0.005 per cent oxygen will have occurred (Machta and Hughes, 1970). Combustion of all known high grade fossil fuel reserves (coal, oil and gas) could reduce atmospheric oxygen levels by 1.8 per cent. The total mobilization of massive low organic carbon deposits in oil shales, however, could deplete the atmospheric pool by 50 per cent of its oxygen. It is highly unlikely that all these resources will ever be economic or technologically possible to recover but dramatic changes in the oil economy of the Industrial West has directed attention to low grade fossil carbon deposits. The high demand for fossil fuel has resulted in a massive technological input for comparatively low energy returns (about 4 MJ kg^{-1} processed material compared with 30 MJ kg^{-1} for bituminous coal). The working of these deposits has also created major environmental problems in the disposal of spent sand and shale. We will see that the mobilization of carbon as atmospheric CO_2 is more likely to place restraints on the use of fossil fuels than the consumption of oxygen.

Approximately 93 per cent of the oxygen and 99 per cent of the total carbon is located in the sedimentary pools, particularly as carbonates, and less than 1 per cent of the total carbon pool is located in the biosphere. This utilized pool is distributed rather unevenly with only about 0.2 per cent in the oceans given that they occupy some 71 per cent of the earth's surface. We can deduce from the ratio of biomass to production (GPP) on land of about 17:1 compared with about 0.07:1 in the oceans that the turnover of hydrogen and oxygen as well as carbon is faster in the marine environment. The residence time for carbon in living terrestrial vegetation is approximately 15 to 20 years compared with less than a month in the small marine plant biomass. The principle reason for this is that the bulk of marine plants are unicellular organisms (phytoplankton) which have a very short life compared with woody terrestrial vegetation. The turnover of dead organic matter is much slower, particularly on land where cellulose and lignin form a major proportion of plant tissues. These materials decompose slowly (see Chapter 7) and there is a general tendency for organic matter to accumulate in soil. Under most conditions a balance is reached between the input of carbon to the decomposer subsystem and its mineralization as CO_2. The reciprocal processes of production and respiration may, however, be separated temporarily. This is shown on a global scale by the seasonal changes in carbon dioxide (CO_2) content of the atmosphere in the Northern Hemisphere. During the summer, fixation of CO_2 exceeds total respiration while the reverse is true of the winter months. The variation is as much as 20 ppm CO_2 and represents an annual gross fixation of about 5.5×10^9 tonnes of carbon. This is about a quarter of the annual terrestrial production and agrees with independent estimates of carbon fixation in the Northern Hemisphere.

The same relative displacement of production and community respiration occurs over 24 hours within ecosystems and the diurnal CO_2 fluxes may be used to estimate primary production (Woodwell, 1970).

The production to respiration ratio is approximately one, both on land and in the oceans, though the rate of turnover of organic materials is rather different. These two largely independent biological cycles are dynamically connected by a much slower exchange of inorganic carbon, as CO_2, between the atmosphere and hydrosphere. The pool of dissolved inorganic carbon is approximately 50 times that of the atmosphere. The balance of gaseous and dissolved CO_2 is held in dynamic equilibrium by a series of reversible reactions:

$$CO_2 + H_2O \rightleftharpoons H_2CO_3 \rightleftharpoons H^+ + HCO_3^- \rightleftharpoons 2H^+ + CO_3^{2-}$$

The direction of the reaction is dependent upon the relative concentrations of the components and a number of other physicochemical factors (Raiswell *et al.*, 1980). A depletion of atmospheric CO_2 by terrestrial plants or a depletion of bicarbonate by phytoplankton triggers a complex series of compensatory reactions. The rate of exchange between the atmosphere and hydrosphere is comparatively slow and has been determined by the removal of elevated ^{14}C isotope levels in the atmosphere produced by atomic bomb tests. The solution of ^{14}C in the oceans and its dilution in the atmosphere by 'cold' carbon ^{12}C indicates an atmospheric residence time of 5 to 10 years and a reciprocal annual flux of some 10^{11} tonnes of CO_2 between the atmosphere and hydrosphere.

This homeostatic process does not, therefore, buffer the seasonal changes in the earth's atmospheric CO_2 concentrations but operates over a longer time span. There is evidence, however, that this long-term homeostasis is not globally stable and that man's activities may have shifted the equilibrium point. Over a period from 1958 to 1976 the mean annual CO_2 concentration in the atmosphere has risen by 0.8 ppm from about 316 ppm to 332 ppm (Raiswell *et al.*, 1980). It is believed that this has principally arisen from the burning of fossil fuels and the destruction of forests. At the present time approximately 5×10^9 tonnes of carbon are added to the atmosphere each year by the burning of fossil fuels. About half of this remains in the atmosphere, half dissolves in the oceans and a small percentage is added to the land biomass (Bacastow and Keeling, 1973). The tropical forests of the world contain some 72 per cent of the world's organic (non fossil) carbon in the form of trees and soil organic matter. These forests are being cleared at a rate of 11 million ha yr^{-1} and it has been calculated that the burning of timber and aeration of the soil has resulted in a release of carbon twice that from fossil fuels. It seems unlikely that this trend will be halted and a wide range of forecasts have been made of atmospheric CO_2 levels by the turn of the century. Concern over this phenomenon centres on the increased absorption of infrared radiation in the atmosphere by CO_2 causing a rise in global temperature as a result of the 'greenhouse' effect. One of the consequences of this could be a mobilization of water stored in the polar ice caps and a reduction in the continental land mass as the level of the seas rise. The difficulties of measuring changes on this scale, with estimates of carbon fluxes or pools varying several orders of magnitude between authorities, prevent general acceptance or dismissal of these predictions. The concern, however, must remain in the face of so much other evidence of global changes in man's environment.

About 6 per cent (dry weight) of the global net primary production is formed by hydrogen atoms split from water. On land, however, the hydrogen incorporated in photosynthetic products is a very small proportion of the total water flux through the plant. The production of 18 tonnes (fresh weight) of cereal requires the removal of 1820 tonnes of water from the soil. Of this 14 tonnes is in transit and only 3 tonnes represents fixed hydrogen (Penman, 1970). The enormous flux of water through the plant results from transpiration; water taken up by the roots carries minerals to the leaves for biosynthesis and is evaporated from the leaf surface. The amount of water transpired by trees is considerable. A Douglas fir plantation, for example, transpires some 48 tonnes of water per hectare per day (Ovington, 1965). The uptake of this water from the soil significantly reduces surface runoff, leaching and erosion. It is predictable, therefore, that the destruction of tropical forests, in some of the world's highest rainfall areas, may have long-term consequences for the water balances of the continents over and above the effects on the carbon cycle. We can see further evidence of the impact of man on biogeochemical cycles in the nitrogen cycle.

The nitrogen cycle

The nitrogen cycle (Fig. 2.7) is the most complex of the nutrient cycles not only because of the large variety of forms in which it is present but also because so many of the steps in the cycle are mediated by specific microorganisms.

The atmosphere is the main exchange pool of nitrogen but it is unavailable in gaseous form to the majority of organisms. Nitrogen principally enters the biosphere through the activities of nitrogen fixing bacteria and blue/green algae. Free-living nitrogen-fixers convert gaseous nitrogen to proteins and the nitrogen becomes available to higher plants after the prokaryotes decompose. Nitrogen fixation is carried out aerobically by members of the Azotobacteriaceae (e.g. *Azotobacter*) and by many Cyanophyceae (blue-green algae or Cyanobacteria). A large number of *Clostridium* and *Desulphovibrio* species, as well as some *Bacillus* and *Enterobacter*, fix nitrogen anaerobically. In symbiotic systems (principally *Rhizobium* in the root nodules of Leguminosae such as clover) N_2 is made available to the host as ammonia (NH_3). Nitrogen in plant protein is utilized by the herbivore subsystem or passes directly to the decomposer subsystem together with animal bodies and excretory products. Ammonification results in the formation of ammonia from amino acids. The

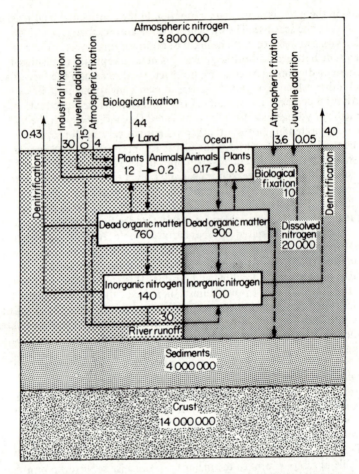

Fig. 2.7 The nitrogen cycle. Distribution of nitrogen in the biosphere and annual transfer rates can be estimated only within broad limits. The two quantities known with high confidence are the amount of nitrogen in the atmosphere and the rate of industrial fixation. The apparent precision in the other figures shown here reflects chiefly an effort to preserve indicated or probable ratios among different inventories. Thus the figures for atmospheric fixation and biological fixation in the oceans could well be off by a factor of 10. The figures for inventories are given in billions of tonnes; the figures for transfer rates are given in millions of tonnes. Because of the extensive use of industrially fixed nitrogen, the amount of nitrogen available to land plants may significantly exceed the nitrogen returned to the atmosphere by denitrifying bacteria in the soil. A portion of this excess fixed nitrogen is ultimately washed into the sea, but it is not included in the figure shown for river runoff. Similarly, the value for oceanic denitrification is no more than a rough estimate that is based on the assumption that the nitrogen cycle was in overall balance before man's intervention. (From Delwiche, 1970.) Sönderlund and Svensson (1976) provide a detailed review of the N cycle which challenges some of the values given here for both pools and flux rates.

dissolved ammonia may be taken up by plants from the soil solution or undergo nitrification so that it is absorbed by the roots as nitrate.

$$NH_4^+ \rightarrow NO_2^- \rightarrow NO_3^-$$

The reverse process of nitrification, de-nitrification, is carried out by bacteria which, under anaerobic conditions use nitrogen oxides as terminal electron acceptors instead of oxygen.

$$NO_3^- \rightarrow NO_2^- \rightarrow \text{intermediates} \quad \overset{\nearrow NH_3}{\searrow N_2O \rightarrow N_2}$$

The activity of denitrifying bacteria which reduce nitrate (NO_3^-) to N_2 (e.g. *Pseudomonas* and *Thiobacillus denitrificans*), N_2O and NH_3 (many bacteria) is of considerable ecological significance since this represents potential losses of nitrogen from the system.

Current estimates of biological nitrogen fixation are in the order of 126 million tonnes a year with spontaneous fixation (through lightning discharges, volcanic activity and ultraviolet radiation) contributing a further 26 million tonnes per year (Postgate and Hill, 1979). Most of the biological fixation occurs on land, principally by symbionts, but the annual uptake of nitrogen by natural vegetation is estimated as 35 times higher than fixation rates (Ehrlich *et al.*, 1977). This indicates both the importance and the scarcity of fixed nitrogen to living organisms since it is conserved and recycled in its available forms within biological systems. Nitrogen, however, shows no evidence of continuous accumulation and immobilization under natural conditions so denitrification rates are assumed to balance fixation rates, except for a small loss of nitrogen to deep sea sediments. Blooms of nitrogen fixing blue-green algae may provide between 2 per cent and 30 per cent of the N input to marine phytoplankton (Postgate and Hill, 1979), but the primary source of N for most aquatic systems is from the land. This view is challenged by Sönderlund and Svensson (1976) who consider that sediments are an unrecognized site of nitrogen fixation.

At the beginning of the twentieth century the principal source of fixed nitrogen for fertilizers was from nitrate deposits in Chile. In 1914 an industrial method, the Haber Process, was developed for the fixation of N_2 gas. Industrial fixation is still carried out by this method and in 1976 amounted to 48 million tonnes per annum. The industrial process requires considerable energy input as well as a hydrogen source. Currently both the energy and the feed are supplied by fossil fuels, principally gas and oil. Nitrogen fertilizers and food prices are thus closely linked to our present fossil fuel reserves. The world demand for food has resulted in an 8–10 per cent increase per annum of nitrogen fertilizer. It has been estimated that a country must build one new Haber plant, producing one tonne of NH_3 per day, for every 6 million people added to the population. India, at its mid-1970 rate of population growth, required the construction of 2½ such plants a year (Postgate and Hill, 1979). Even if world demand for nitrogen fertilizer could be equated with technological developments and declining oil reserves, the environmental costs of such an escalation would probably be unacceptable. Only one-third or less of the fertilizer applied to soils is taken up by agricultural crops and the remainder is leached to freshwater and marine systems. Much of the nitrogen excreted by human populations also finds its way into aquatic systems. Denitrification has not generally compensated for the overloading of aquatic systems with nitrogen, and eutrophication and nitrate contamination of drinking water are well known phenomena. In addition there is well documented evidence that carcinogenic nitrosamines are accumulating in the environment and that nitrous oxide, produced as a by-product of bacterial denitrification, may be having deleterious effects on the ozone layer in the atmosphere (Arrhenius, 1977).

Both economic and environmental concerns have focussed attention on biological nitrogen fixation (q.v. Postgate and Hill, 1979). Microorganisms fix N_2 at a fraction of the industrial energy cost using an enzyme nitrogenase; the total biological fixation of N_2 in the biosphere is probably mediated by no more than a few kilograms of this enzyme. While biological fixation of nitrogen offers a possible solution to the increasing world demand for fertilizer, it does not offer solutions to the problems of cultural eutrophication and the overall apparent imbalance between fixation and denitrification.

Sedimentary Cycles

Phosphorus
Phosphorus is essential to living organisms but although it is required in only about 10 per cent of the quantities of nitrogen, it is more frequently a limiting nutrient for plant growth. This primarily results from the absence of a gaseous phase of the element and a predominance of pathways linked to the geologic cycle (Fig. 2.8).

Some fraction of the phosphorus leached from land is deposited in deep ocean sediments where it is effectively lost on a human time scale since diagenesis probably takes 10^8 to 10^9 years. Thus the average concentration of phosphorus in sedimentary rock is lower than that in primary igneous rock and as the geological cycle proceeds the continental masses, and hence the biosphere, become progressively deficient in this element.

Phosphorus is made available to higher plants as inorganic phosphate (PO_4^{3-}) which is released from rocks by weathering or by the decomposition of organic matter by the decomposer subsystem. Phosphorus is efficiently cycled and conserved by living systems but on each bio-

Land   Ocean

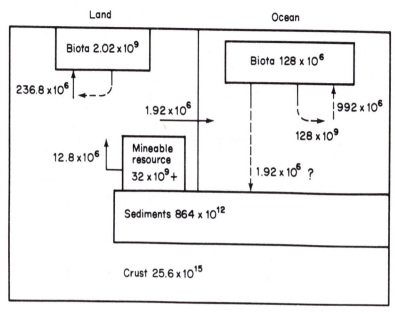

Fig. 2.8 The global phosphorus cycle. The figures include the effects of industry and agriculture on the size of phosphorus pools and fluxes. Values are in tonnes per annum. (After Pierrou (1976); from Stumm, W. (1973). *Water Research*, 7, 131–44.)

logical cycle a small proportion of the element is lost to the sedimentary (reserve) pool. Phosphorus salts may become so tightly bound to soil complexes of iron, aluminium, calcium and clays that they are unavailable to plants.

Less chemical immobilization of phosphate ions occurs in aquatic habitats than soils because of lower concentrations of iron, aluminium and calcium salts; though phosphorus may become physically unavailable to plants as a result of deep waters becoming thermally stratified (p. 52). High concentrations of available phosphorus, from industrial wastes, sewage and detergents, discharged into water bodies usually lead to excessive plant growth (eutrophication), particularly unicellular algae. When the algal bloom has depleted the available nitrogen in the water, nitrogen fixing blue-green algae become dominant and can result in the accumulation of metabolites which are toxic to cattle, fish and water birds (Duffus, 1980). Algal growth continues until the phosphorus supply is exhausted or other minerals become limiting; the ecosystem may or may not then recover its original community structure. Phosphorus is generally accepted as the primary agent of eutrophication in most aquatic ecosystems.

The turnover of phosphorus by living organisms may be extremely rapid, ranging from as little as 10 minutes in the surface regions of freshwater lakes to several hours in the oceans (Pomeroy, 1970). The turnover in soils is much slower and Reiners and Reiners (1970) have suggested that it may be in the order of 200 years in organic soils.

Under natural conditions phosphorus is transferred from the lithosphere to the hydrosphere bound to mineral particles, and not predominantly in soluble form. The loss of phosphorus from the land is therefore closely linked with other physical, chemical and biological phenomena which affect erosion rates.

As a consequence of its scarcity and its demand for food production, man has probably had a greater gross effect on the phosphorus cycle than that of any other element. Worldwide mining of phosphate rock amounted to 94 million tonnes in 1970, of which 70–80 per cent was added to the land as fertilizer; detergents, animal feed additives, pesticides and medicines account for the remainder. The major trend is the disposal of phosphorus from a concentrated form in mineral deposits to non-recoverable concentrations in aquatic sediments. The major transfer pathway is the disposal of domestic sewage in aquatic systems. The total returns of phosphorus from the hydrosphere to the lithosphere (as guano and fish) are less than a hundredth of the flux in the reverse direction and do little to redress this imbalance. In the long term phosphorus could become limiting to agricultural productivity but short-term concerns are principally directed towards the effects of eutrophication on drinking water resources, conservation and freshwater fisheries. The global phosphorus cycle is discussed in detail by Pierrou (1976).

Other mineral cycles
The nutrients considered above represent only five out of the 21 elements in Table 2.2. All of these have particular characteristics of flow and cycle but all except sulphur, which has important gaseous phases, are linked to sedimentary pathways. Various elements such as calcium, magnesium, potassium, silicon and even vanadium are required in significant amounts of particular organisms – calcium in the shells of molluscs, magnesium for the chlorophyll of green plants, silicon for the tests of diatoms and vanadium for the respiratory pigments of tunicates (sea squirts). A wide range of trace elements also cycle naturally through the biosphere (Table 2.2). Some of these are essential micronutrients at low concentrations but toxic at higher levels, others have no known biological function. Among this group are the elements of greatest environmental concern (lead, mercury, cadmium, arsenic, berillium, nickel, chromium, selenium, vanadium, molybdenum, copper and zinc) which have a wide spectrum of toxicological effects (Duffus, 1980).

Plant species vary considerably in their tolerance of heavy metal concentrations in soils; a phenomenon which is of considerable importance in the reclamation of mine waste (Bradshaw and Chadwick, 1980) and, paradoxically, may also be used to locate new mineral resources. Warren (1972)

Table 2.3 Comparison of flows of trace metals for human activities and natural processes (t yr^{-1}). (From Ehrlich *et al.*, 1977.)

Element	Mining	Emission to air	Natural rainout	Natural stream load
Zinc	5000	730	1000	370
Lead	3000	400	310	180
Chromium	2000	50	120	7
Arsenic	60	50	190	15
Cadmium	14	4	NA	<40
Mercury	9	10	1	3

(NA = not available)

describes the technique of 'biogeochemical prospecting' which he pioneered to overcome the costly and often dangerous labour involved in prospectors removing overburden from the soil surface to sample the underlying rock for ores. The idea was conceived that the metal content of some plants might reflect the nature of the underlying strata. Results for copper were highly successful. The ash of Blueberry (*Vaccinium* spp.) leaves at one Canadian site contained 2000 ppm copper compared with 570 ppm in an uncontaminated adjacent area, while Green Alder (*Alnus sinuata*) concentrated even higher levels of copper at 25 650 ppm and 310 ppm respectively. The technique also showed promise for locating zinc, molybdenum, mercury, gold and silver but in all cases careful biological interpretation of the soil/plant relationships was required. Some animal groups, particularly crustacea and molluscs, also have an affinity for accumulating massive body loads of heavy metals and can be used as a means of detecting and monitoring sources of cadmium, copper and zinc pollution (Stenner and Nickless, 1974).

Over geological time scales surface rocks containing heavy metals have weathered to a point where the natural soil or stream load is generally low. Man's industrial uses of these elements and/or their presence as contaminants in ores and fossil fuels has resulted in increased mobilization to levels equal to or exceeding natural rates (Table 2.3). This mobilization is the result of the exposure of low grade ore and spoil to primary weathering and losses in processing and handling. Sulphur is particularly important in this respect as man has doubled natural sulphur fluxes not only through his use of coal and oil but also through a very appreciable influence on weathering rates (Fig. 2.9). Sulphur emissions as sulphur dioxide (SO_2) into the atmosphere are distributed very unevenly over the globe. In N.W. Europe 80 per cent of the total SO_2 is anthropogenic and the acidification of rain by solution of the gas is a major source of environmental concern (p. 37).

Heavy metal resources are finite and non-renewable. Their mobilization into dispersed forms, which are not economic or practical to recover, has

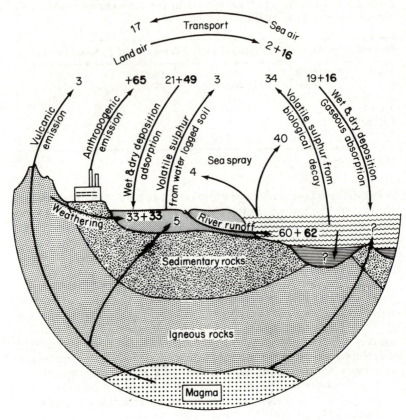

Fig. 2.9   The global sulphur cycle. Transfers are shown in 10^6 t yr^{-1}; estimated transfers before man significantly affected the cycle, with anthropogenic effects in bold type. Note that the transfer of sulphur from sea to land before industrialization is now reversed (e.g. the net atmospheric flux of sulphur was 15 t yr^{-1} from sea to land before intensive industrial development took place, whereas the present net flux is 1 t yr^{-1} from land to sea). (From Granat, L. *et al.* (1976). *Ecol. Bull. (Stockholm)*, **22**, 89–134.)

long-term significance for biological systems as well as man's industrial economy. Lead is particularly well documented because the cycle is both anthropogenic (produced by man) and the effects are anthropocentric (primarily affecting man). Unlike some of the other heavy metal pollutants, such as copper, manganese, zinc and cadmium, lead affects man directly. The effects are severe where man is exposed to high concentrations of the element in the urban and industrial environment.

The earth's known exploitable resources of lead (principally as the sulphide ore galena) are about 100 million tonnes. Natural weathering and

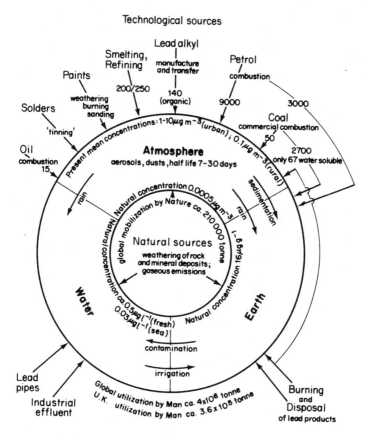

Fig. 2.10 Sources of lead in the environment. Values are in t yr^{-1}. (Reproduced with permission from *Lead in the Environment and its Significance to Man.* (1974). H.M.S.O., London.)

volcanic activity releases about 210 000 t yr^{-1} into the biosphere and natural levels in soils (away from mineral deposits) range from 2—200 ppm. Lead salts are generally rather insoluble and concentrations in aquatic systems one or two orders of magnitude below soils are fairly general.

In comparison with the natural mobilization rates of lead, man's present usage is high (Fig. 2.10) amounting to about 4 million tonnes per year. A significant feature of the use (and scarcity) of lead is that about half of it is recycled as metal in the form of battery casings, cables, pipes and foil. The other 50 per cent is not recoverable and is dispersed, principally as aerosols, from antiknock additives to petrol ($12\,000$ t yr^{-1}), coal combustion (2750 t yr^{-1}), smelting and refining ($200-250$ t yr^{-1}). These aerosols have a half-life in the atmosphere of 7 to 30 days before they enter terrestrial or aquatic systems. A comparatively small amount of lead is mobilized in soluble form as lead salts have low solubility except in acid waters. The principle mode of uptake by animals is in the form of aerosols or dust on vegetation. Urban man is at greatest risk. The concentration of lead in city dusts often lies in the range 1000–4000 ppm with values as high as 50 000 ppm (5000 ppm indoors) near lead works. High lead concentrations in soils and dusts are particularly hazardous since food plants may absorb lead, dusts settle out on foods and children playing in dusty areas may ingest large quantities of lead. Lead poisoning in children commonly results from the ingestion of lead-based paints.

Man has been exposed to lead poisoning for centuries; it is even suggested that it may have hastened the decline of the ancient Greek and Roman civilizations. Only two generations ago (with a population of about a third of the present level) more than 1000 people in the U.K. suffered annually from industrial lead poisoning and over 30 died. Today about 70 cases of industrial poisoning occur each year with the last recorded death 18 years ago. Children are at greater risk and approximately 100 are admitted to hospital in the U.K. each year, with about two fatalities directly attributable to lead poisoning.

Wild animal populations are only exposed to high levels of lead in the urban environment or on roadside verges. Studies on lead levels in small mammals have shown that the concentration is related to distance from the road and traffic levels. There is no evidence that levels as high as 22.7 ppm (dry weight) significantly affect small mammal population densities (Quarles *et al.*, 1974).

Lead is the most abundant but fortunately not one of the most toxic heavy metals dispersed into the biosphere. The main source of poisoning is by direct ingestion but other metals such as cadmium and mercury are more biologically active. These metals, as well as some pesticides and organic industrial compounds, become concentrated in animal tissues due to transfer from plants or animal foods (p. 93). In order to appreciate these effects it is necessary to understand the structure of ecosystems and the subsystems they contain.

3 Ecosystems

Climate (essentially temperature and rainfall) is the main factor governing the distribution and abundance of organisms on land, while light and nutrient availability are generally more important in aquatic systems. Within the biosphere we can recognize particular regions which have characteristic climatic, or other physical conditions, which support a particular flora and fauna. These regions are the *biomes*. The oceans may be considered as a single biome although subdivisions into the continental shelves, coral reefs and deep oceanic regions may be recognizable. It is inappropriate to consider these oceanic subsystems as biomes because of the high degree of interdependence which exists between them in terms of nutrient supply and cycling.

The Terrestrial Biomes

The general relationship between climate and the major terrestrial biomes is shown in Fig. 3.1. The overall pattern is one of forests dominating high rainfall areas with forest biomass decreasing from the equator to the cold temperate regions as the mean annual temperature decreases. Under lower rainfall conditions in the same region grasslands are generally established. Under very low levels of precipitation deserts are formed in tropical, temperate and subpolar regions. All three of these major biome types effectively converge in the tundra where areas of stony deserts, grass/ moss meadows and dwarf shrubby vegetation are found according to local soil and moisture conditions.

The *tundra* extends around the Arctic Circle south of the permanent ice cap and north of the tree line. The terrain is generally flat, as a result of repeated glaciation. Subzero temperatures dominate the climate and the soil is permanently frozen below the surface layers (permafrost). A mozaic of shallow pools is formed over low-lying ground during the short summer period of two months or less. Trees are absent and the vegetation consists of a low mat of grasses, sedges, moss, lichens and dwarf shrubs. The low temperatures of the growing season limit transpiration so that water availability is not a major factor governing the distribution of plants. Reindeer (in Europe) or caribou and musk ox (in America) and lemmings are the main grazing animals. Large populations of birds

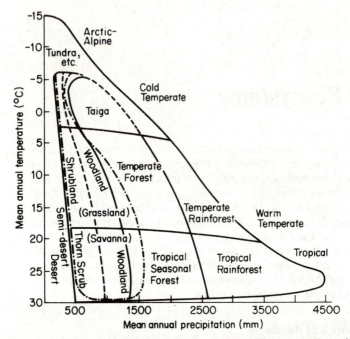

Fig. 3.1 A pattern of world biome types in relation to climatic humidity and temperature. Boundaries between types are, for a number of reasons, approximate – in climates between forest and desert the balance between woodland, shrubland and grassland types may be shifted due to maritime versus continental climates, soil effects and fire effects. The dot-and-dash line encloses a wide range of environments in which either grassland, or one of the types dominated by woody plants, may form the prevailing vegetation in different areas. (From Whittaker, 1975.)

migrate to the tundra to breed. The continuous daylight hours and vast swarms of insects promote rapid growth of the young during the short summer season. Increasing exploitation of low grade oil deposits (oil sands and shales) is having a considerable impact on the tundra since the slow growth rate of plants limits the rate of recovery from damage.

Taiga or *boreal forest* is found in a broad belt across Eurasia and North America, bounded in the north by tundra and merging into mixed forest types in the south. The evergreen boreal forest is composed largely of spruce, pine and fir which can tolerate the long cold winters and photosynthesize whenever temperatures rise above freezing. The resin content of evergreens makes them unsuitable food for most insect and vertebrate herbivores. Moose, snowshoe hares and birds are generally associated with the forest borders and lakes where grasses and deciduous shrubs, such as willows, are found. Lynx, foxes, bears, wolves and birds of prey are the main predators.

In temperate regions we can recognize two main biome types whose distribution is largely determined by water availability. *Temperate deciduous forest* is predominantly found in Western Europe, Eastern U.S.A., China, Japan and to a limited extent in the southern parts of South America, Africa and Australasian continents. Broad-leafed trees such as oaks, beeches and maples, which shed their leaves before winter, are dominant. Winters are often cold but warm, wet summers produce a generally humid climate with a long growing season and long periods of drought are rare. Insects, birds and mammals are abundant. The larger mammals include rabbits, squirrels, deer and wild bear with their predators the foxes, wild cats, bears and the mountain lion.

The *temperate grasslands* occur in regions with dry summers and low winter temperatures. These features combine to produce a summer soil water deficit which excludes the potential for tree growth except around lakes and watercourses. The prairies of North America and the steppes of Asia are the principal regions of this biome but there are more limited areas of temperate grassland in the Southern Hemisphere, notably the pampas of South America and the veldt of Southern Africa. The large mammals, with the exception of the North American bison, are not well represented in natural grasslands in comparison with tropical savannah. The principal vertebrate herbivores in temperate grasslands are small rodents (voles, mice and ground squirrels).

The *savannah* areas of Africa, South America and Australia have a long, dry season with a soil water deficit for at least 5 months. The resulting vegetation is a tropical grassland with scattered trees. The density of the trees varies according to rainfall and merges into seasonal forest in the moister regions. In modern times much savannah is burnt annually and large areas are maintained as grasslands where forest would normally develop. The African savannahs support the most varied vertebrate fauna in the world.

The *tropical forests* extend through the equatorial zone in a band covering West Africa, Malaysia, Burma, Indonesia, North East Australia and a major proportion of South America. At the latitudinal margins of its distribution the climate is seasonal with a short, dry period limiting production and decomposition. The rain forests are found in areas of high rainfall, distributed throughout the year, with a temperature regime which shows little seasonal variation. The trees are often tall, some may exceed 50 m, and many species have buttressed bases. There is a luxuriant flora of epiphytes (plants growing on or supported by the trees) and most species, particularly the lianas, flower high up in the forest canopy. In most mature rain forests little light penetrates through the canopy and the undergrowth is usually sparse. Plant and animal diversity reaches its maximum in this biome though animal biomass, particularly of herbivores, is much lower than in the savannah. Mammals, other than primates, are poorly represented in rain forests and are primarily arboreal, together with the birds and reptiles. Insects, particularly termites, are abundant.

A more detailed description of the distribution and classification of the terrestrial biomes can be found in Walter (1973).

The Watershed Approach to Ecosystem Functioning

Within the terrestrial biomes we can recognize a number of ecosystem types which are not usually defined by sharp boundaries. This can present problems in the empirical study of ecosystems, particularly where transfer of nutrients from land to water is a significant aspect of functioning and integration of adjacent ecosystem types. For this reason it is often appropriate to consider the catchment area (or watershed) of a stream or lake as the appropriate ecological unit. The value of this approach is that the functioning of whole systems can be measured in terms of nutrient inputs and outputs across the boundaries without detailed consideration of the interacting subsystems and their components. Ultimately, of course, the internal structure has to be analysed to interpret the functioning of a natural system or its response to imposed perturbations.

There have been few detailed catchment studies but the most complete and extensive are those for the Hubbard Brook Experimental Forest in New Hampshire, U.S.A. (Whittaker *et al.*, 1974, Likens *et al.*, 1970, 1978).

The Hubbard Brook watershed is a mountainous area of deciduous forest ranging from 229 m at the lowest point to 1015 m at the highest point on the surrounding mountain ridge. The underlying rock is largely impermeable to water and a number of small streams drain the dissected slopes of the watershed. Water leaving each of these watershed compartments flows over a V-notch weir so that the volume of water and some of the dissolved nutrients can be measured. The amounts of rainfall, and the nutrients contained in rain water, are also measured so that the net amounts of water and nutrients leaving the watershed each year can be calculated (Table 3.1).

Table 3.1 Nutrient budgets for watershed at Hubbard Brook. Values are in hg ha^{-1} yr^{-1} (Likens *et al.*, 1970).

Element	Precipitation input	*Forested watershed* Stream-flow output	Difference (net output)	*Cut watershed* Difference (net output)
Calcium	2.6	11.7	9.1	77.9
Sodium	1.5	6.8	5.3	15.4
Magnesium	0.7	2.8	2.1	15.6
Potassium	1.1	1.7	0.6	30.4
NH_4 −nitrogen	2.1	0.3	−1.8	1.6
NO_3 −nitrogen	3.7	2.0	−1.7	114.0
Sulphur	12.7	16.2	3.5	2.8
Silicon	trace	16.4	16.4	30.0
Aluminium	trace	1.8	1.8	20.7

Calcium, sodium, magnesium, potassium, silicon and aluminium show a net loss from the system indicating that there is a source of the elements within the ecosystem boundary. The source is the bedrock and soil minerals (granite and glacial deposits) which are weathering at a rate of 70 gm^{-2} yr^{-1}, approximately equivalent to a lowering of the rock surface by 0.04 mm yr^{-1}. This rate is near average for land surfaces (Whittaker, 1975) but is lower than on some mountain slopes which are not protected from erosion by the forest cover and particularly the root mat. The homeostasis of the system is emphasized by observations of seasonal stream flow and transported material in dissolved or particulate form. The stream water showed relatively constant concentrations of dissolved minerals despite major changes in output after storms or snow melt. These seasonal changes in current velocity also had little effect on the erosion of particulate matter and over the year output of suspended solids was smaller than dissolved solids. The difference between precipitation (1246 mm m^{-2} yr^{-1}) and run-off (799 mm m^{-2} yr^{-1}) of 447 mm m^{-2} yr^{-1}, or 36 per cent of precipitation, provides a measure of the total evaporation and transpiration of the forest system (evapotranspiration).

The pH of rain and snow at Hubbard Brook was often below 4.0, and the sulphate concentration was as high as 5 mg 1^{-1}. Both these features are suggestive of industrial pollution of the atmosphere. It was calculated that 50 per cent of the incoming sulphur may have originated from the burning of fossil fuels. This phenomen of 'acid rain' is of widespread concern as it accentuates the leaching of nutrients from leaves by rain water, is toxic to some plants, particularly pines, and accelerates nutrient losses from soils. Clay minerals are negatively charged and absorb mineral cations onto their surfaces. The H^+ in the acid rain water may exchange with these cations and lead to increased losses of minerals from the system.

Inorganic nitrogen differed from the other nutrients in that the precipitation input was larger than the stream-flow output (Table 3.1). Nitrogen will also have been added to the system by fixation and lost by denitrification. We know, however, from our considerations of the nitrogen cycle that nitrogen constitutes a significant fraction of plant biomass. The nitrogen fluxes across the boundary of the system appear, therefore, to be negligible in proportion to the internal nutrient cycle although we have no direct evidence of this from the data since the ecosystem is effectively a 'black box', i.e. the internal organization is unknown. This question could be resolved either by investigating the internal structure or by perturbing the system and observing changes in the outputs. Both of these approaches were, in fact, carried out simultaneously on different areas of the watershed but the data in the final column of Table 3.1 show the dramatic effects of deforestation followed by herbicide application to stop regrowth.

The effects of this major perturbation to the forest were an increase in stream flow by 41 per cent, 28 per cent and 26 per cent over three years (an increase in runoff approximately equal to the calculated evapotranspiration) and a massive export of mineral nutrients. Calcium output increased by 417 per cent, magnesium by 408 per cent, potassium by 1558 per cent and nitrogen by 5600 per cent! Thus large pools of these nutrients, which had been accumulated in soil and biota within the system during the course of ecological succession, were mobilized by the perturbation. The increased loss of elements could not be simply explained in tonnes of the increased water discharge because of the disproportionality of the effects. The basic reason for this is obvious: the deforestation has broken the internal cycle between the autotroph and decomposer subsystems. The mechanisms for the mobilization of individual nutrients, however, are more complex and must be considered with respect to the internal structure of the system.

The Structure and Functioning of Ecosystems

The empirical analysis of ecosystem structure requires the delimitation of the boundaries. Sometimes these are defined naturally but in a region of more or less homogeneous terrestrial vegetation, or on a vegetation gradient, the selection of these limits is more subjective. Within the International Biological Programme* it was conventional to delimit a type hectare, though the size of the type depended upon vegetation homogeneity. Only non-destructive measurements, such as temperature, rainfall or CO_2 fluxes, were made within this area. The effects of trampling, a very significant factor in delicate systems such as the tundra, were confined to strictly prescribed paths or board walks. The type hectare was surrounded by an undisturbed buffer zone, the importance of which, again, should not be underestimated in any ecological studies. Around the outside of this zone sampling areas were delimited ranging from 1 ha to 10 ha for vegetation and soil analyses to much larger areas for mammal and bird studies.

The definition of an ecosystem, within this programme of research, was an area which is self-contained in terms of its primary production and nutrient cycling. That is to say, the extent of nutrient fluxes across the boundary was small with respect to the internal pathways. It is possible to recognize a pond, lake or the oceans as falling within this definition but flowing water systems tend to have a major through-flux of nutrients and little internal production and cycling. Streams and rivers may therefore be regarded as open ecosystems or more strictly, like the soil, as a subsystem of a larger complex (see Fig. 3.4). The precise definition of the ecosystem is thus open to debate but the principle of the interacting sub-components of the biological system is the essential nature of the concept.

*An international co-operative study of ecosystem structure and functioning in the major biomes conducted from approximately 1967–1975.

This holistic concept of ecosystem structure and functioning was developed through the work of a number of eminent ecologists in the early part of this century. Charles Elton (1927), for example, described the biotic structure of ecosystems by allocating the organisms to a number of broad trophic levels (plants, herbivores and carnivores). A graphical presentation of the trophic structure of most ecosystems results in pyramids of numbers or biomass reflecting the decreasing abundance of organisms at higher trophic levels, e.g. foxes < mice < grass on a biomass or density basis per unit area. Subsequently, in 1942, Raymond Lindeman drew attention to the quantitative relationships which exist between these different trophic levels in terms of energy and matter transfers which maintain the functional integrity of the system (see p. 38). This 'trophic dynamic' concept, as it was called, stimulated much of the quantitative analysis of ecosystems which is fundamental to our present understanding of this aspect of ecology.

A generalized model of ecosystem structure is shown in Fig. 3.2 illustrating the three subsystems, plant, herbivore and decomposition, and their main compartments or trophic levels. The functioning of the ecosystem may be described in terms of energy or nutrient fluxes but both approaches

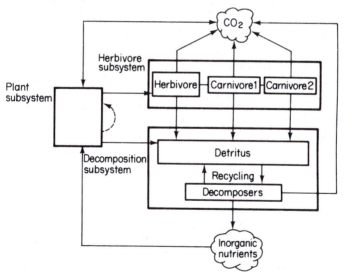

Fig. 3.2 A general model of ecosystem structure. The three subsystems, plant, herbivore and decomposition, are shown together with their main components. The major pathways of transfer of matter within the ecosystem are shown by the arrows. Organic matter pools are shown as rectangles, inorganic as 'clouds'. Note in particular the links between the herbivore and decomposition subsystems, the recycling of matter *within* the decomposition subsystem and the net storage of matter that may occur within the plant subsystem (dashed arrow). (From Swift *et al.*, 1979.)

have their value and limitations. Ecosystem energetics (as well as at community and population levels) are of particular value for comparing different systems since plant and animal trophic levels, and their interactions, can be expressed and compared in a single unit of energy, the joule or calorie (1J = 0.24 cal). In many situations where the availability of mineral nutrients, rather than energy, limits the rate at which the system functions, the investigation of nutrient flux pathways may be more meaningful. Mineral nutrient budgets also emphasize the integrity of the ecosystem and the feedback loops operating within the boundaries which regulate and maintain this functional integrity.

There have been few complete studies of energy flux through ecosystems but two classical examples are shown in Fig. 3.3.

Fig. 3.3(b), a forest ecosystem, emphasizes the large stored energy content of the plant biomass in comparison with the gross primary production. The energetic cost of maintaining this biomass is low due to the predominance of metabolically inert woody tissues. The herbivore biomass is small and the main energy fluxes are autotroph and decomposer respiration.

The marine bay plankton community (Fig. 3.3(a)) has, by comparison, a minute autotroph biomass supporting a proportionately higher energy flux through the larger herbivore subsystem. This phenomenon is characteristic of planktonic communities and is known as an inverted biomass pyramid.

Carbon and energy are closely linked in the biosphere, and ecosystem structure and functioning can be similarly described in terms of carbon fluxes. Carbon flow in the Hubbard Brook system (see p. 36) is shown in Fig. 3.4 and is a good illustration of the functional integration of adjacent terrestrial and aquatic ecosystems. Both of the types of energy budgets considered above can be seen as integral components of the larger system. The open nature of the stream system should also be noted with respect to its definition as an ecosystem.

The following generalizations can be made on the fate of net primary production in natural ecosystems.

(1)   In terrestrial ecosystems part of the NPP is stored in perennial tissues and contributes to biomass. In grass dominated or aquatic ecosystems this increment is usually negligible but in immature forests or plantations it may contribute 20–60 per cent of NPP.

(2)   A minor proportion of terrestrial NPP is consumed by herbivores. Even in intensively grazed grasslands this fraction rarely exceeds 25 per cent of the total NPP because, although more than half of the above ground biomass may be removed, the annual production of roots may be equal to or exceed above ground production. In forest systems insects or browsing animals rarely consume more than 10 per cent of the NPP, while in aquatic systems 80 per cent or more production by phytoplankton may be consumed by herbivorous zooplankton.

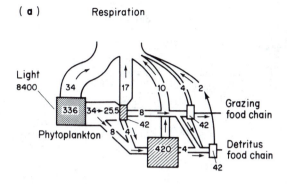

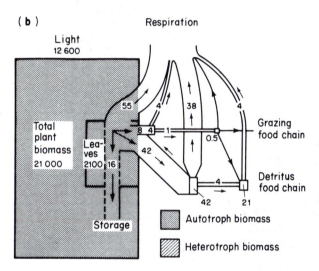

Fig. 3.3 Energy flow through two contrasting types of ecosystem; **(a)** marine bay and **(b)** forest ecosystem. Energy content of animal and plant biomass is in J m⁻² and energy flows in J m⁻² day⁻¹. (After Phillipson (1966); adapted from Odum, E.P. (1962). *Jap. J. Ecol.*, 12, 108–18.)

(3) Material not allocated to plant growth or the herbivore subsystem, together with the remains of the herbivores, predators and excretory products, enters the decomposer subsystem. In ecosystems such as mature forests, which no longer continue to accumulate biomass, a major proportion of net production is shed as plant litter and is processed by the decomposers. Fungi and bacteria are the main agents of decomposition but animals which feed on detritus, fungi and bacteria, as well as their predators, are important components of this community. Material is recycled between the decomposer organisms until carbon mineralization is complete.

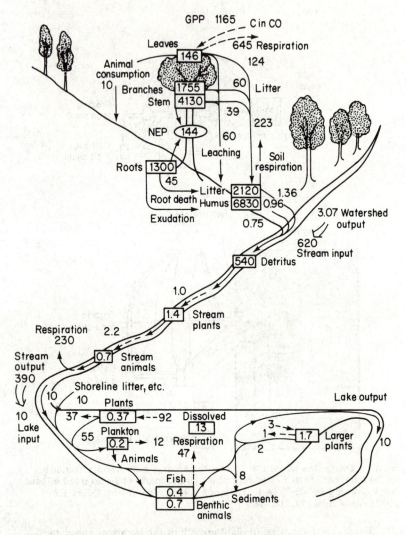

Fig. 3.4 Carbon flow in a landscape at Hubbard Brook from a forest, through a small stream (Bear Brook), to a lake (Mirror Lake). Numbers in boxes are biomass pools in grams of carbon per square metre; numbers on arrows are transfers in gC m^{-2} yr^{-1}. Plant respiration values are estimates. The hollow areas indicate conversion to different area bases when carbon moves from one ecosystem to another: from the watershed (130 ha) to the stream (0.59 ha) to the lake (85.1 ha). (From Whittaker, 1975.)

The following sections briefly examine the gross internal structure and functioning of a forest system, an agricultural grassland and freshwater systems. Emphasis is placed on the pools and flux pathways of nitrogen because they emphasize the feedback loops operating within the systems better than energy, carbon or the non-limiting nutrients which have principally been reported in the literature.

Temperate deciduous forest

The Hubbard Brook Forest is in a relatively immature stage of growth as a result of logging between 1909 and 1917. The total biomass is 16.1 kg m^{-2}, of which 13.3 kg m^{-2} is above ground and 2.8 kg m^{-2} of root biomass. Mature deciduous forests in that region of the southern United States would attain a total biomass of 30–40 kg m^{-2}. The net primary production for the period 1961 to 1969 was estimated as 760 gm^{-2} yr^{-1} above ground and 140 gm^{-2} below ground. The total NPP of 800 gm^{-2} yr^{-1} is low for this type of forest (see Table 5.2). Of the total NPP in this forest system 217 gm^{-2} fell, on average, as woody litter, 273 gm^{-2} as leaf litter and about 25 per cent of the production was accumulated as a growth increment. In a mature forest this growth increment would be negligible. The continued development of this forest ecosystem means that positive feedback loops are operating in the homeostasis of essential nutrients and the nutrient capital is still accumulating. We have already deduced that nitrogen fluxes within the ecosystem boundary must be much larger than the input and output fluxes. This is confirmèd by the nitrogen budget for the system shown in Fig. 3.5. The hydrologic output of nitrogen is less than 0.01 per cent of the total standing crop, which amounts to 5258 kg ha^{-1} (526 gm^{-2}). The nitrogen budget also shows that the net gain of nitrogen from precipitation is supplemented by a nitrogen fixation rate of about 14 kg ha^{-1} yr^{-1} (1.4 gm^{-2}). This value is low in comparison with non-symbiotic nitrogen fixation in grasslands. The reason for this is that the decomposing tree leaves form an acid soil (pH 5.1) and nitrogen fixation is inhibited below about pH 5.5. Denitrification has not been measured.

The nitrogen in plant biomass represents only about 9.5 per cent of the total standing crop. Most of the remainder (90 per cent) is located in mineral soil and soil organic matter with only 0.5 per cent free in the soil solution as ammonium (NH_4^+). A similar distribution of nitrogen is found in evergreen forests as well as tundra ecosystems. In these regions decomposition rates are limited by temperature, litter breakdown consequently takes several years and organic soils tend to develop. Microorganisms decomposing plant remains, particularly fungi, accumulate (immobilize) nitrogen but the turnover of senescent microbial tissues is often slow as a result of low levels of soil animal activity and lysis. Ausmus *et al.* (1976) have shown, for another deciduous forest ecosystem in

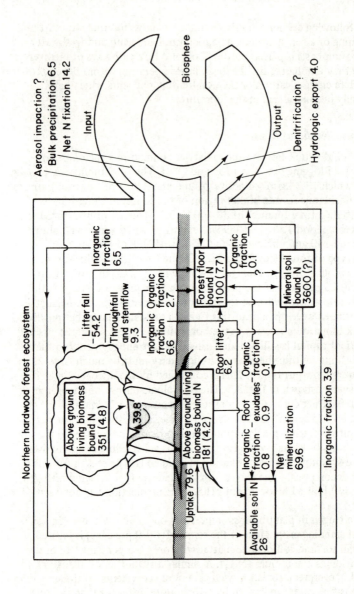

Fig. 3.5 Annual nitrogen budget for an undisturbed northern hardwood forest ecosystem at Hubbard Brook. The values in boxes are in kg N ha^{-1}. The rate of accretion of each pool (in parentheses) and all transfer rates are expressed in kg N ha^{-1} yr^{-1}. (From Bormann, F.H., Likens, G.E. and Melillo, J. (1977). *Science*, **196**, 981–3. Copyright 1977 by the American Association for the Advancement of Science.)

Tennessee, that seasonal shifts can be detected in the location of nitrogen in litter, microbial, soil animal and tree root pools. Microbial immobilization of nitrogen was high during the summer and autumn months but decreased to a minimal level in spring when root growth was at a maximum. The release of nitrogen was attributable to the regulatory effects of soil animal feeding activities on microbial tissues but the transfer of soluble nitrogen from one compartment to another occurred at the peak rainfall period with low leachate losses. The ammonium ion has a positive charge and is attracted to soil colloids. This reduces the possibility of loss from the system in this form. Nitrate (NO_3^-) has no such stabilizing effect and is susceptible to leaching. Under normal conditions in acid woodland soils nitrification appears to be inhibited but increased nitrification, associated with changes in soil redox reactions, following deforestation is the main reason for the huge losses of nitrogen from the Hubbard Brook watershed. The nutrient homeostasis between the autotroph and decomposer subsystems is further integrated by a symbiotic association between plant roots and certain fungi known as mycorrhizae. These fungi do not carry out active decomposition of litter but ramify between the organic particles absorbing nutrients in the mycelium. The absorbed nutrients are translocated to the growing root or stored during the winter period. In return the fungus is supported by the photosynthetic products of the plant. This association is so important in nutrient poor soils that pines and heathers (Ericaceae) will not grow without the mycorrhizal infection.

The location of a major proportion of the total nutrient capital in the soil is shown by other nutrients, such as phosphorus, potassium and calcium, where they are present in suboptimal amounts. (If present in excess they are not retained and will be leached from the system.) In tropical forests decomposition rates are usually high and organic soils do not normally develop (see p. 149). Under these conditions the soil nutrient pools are small, turnover is rapid and 80–90 per cent of the mineral nutrient capital is located in plant biomass. This biomass nutrient pool is more labile than the intractable organic pools in temperate soils. The effect of deforestation on a tropical forest can therefore be more severe in terms of nutrient losses and erosion than in a temperate forest ecosystem.

Nutrient homeostasis also occurs within the autotroph subsystem. Most higher plants withdraw mineral nutrients from senescent tissues before they fall. Woody tissues also have low mineral nutrient content. The flux pathways of nutrients such as nitrogen and phosphorus via litter fall and leaching by rainwater are therefore usually small with respect to internal cycling in the plants. In Fig. 3.5 we can see that the above and below ground tree biomass contains 532 kg N ha^{-1} and that only about 70 kg N ha^{-1} is mobilized through leachates and litter input to the soil. Nitrogen frequently limits plant growth but any one of the other macro or micronutrients can also impose significant restraints on NPP.

(**a**) *Quercus robur* aged 47 years

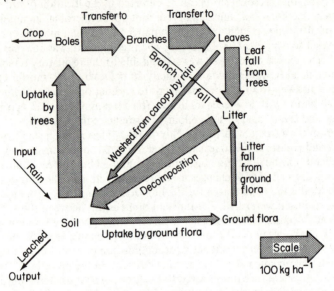

(**b**) *Pinus sylvetris* aged 47 years

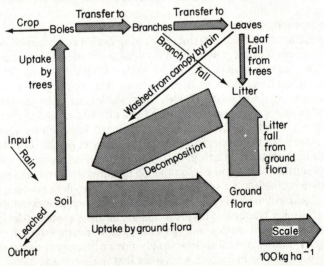

Fig. 3.6 Potassium circulation in adjacent woodlands; (**a**) oak and (**b**) pine, growing under similar conditions. The thickness of the arrow indicates the magnitude of flow. Note the relatively large amount of potassium circulating through the bracken ground flora of the pine plantation and the greater uptake of potassium by the oak trees. (After Ovington, 1965.)

Different plant species also differ markedly in their nutrient requirements and hence nutrient flux pathways in different types of deciduous woodland may vary considerably. Potassium fluxes in adjacent stands of pine and oak woodland are shown in Fig. 3.6. In the oak plantation potassium flow is largely through the oak trees and returned to the soil by leaf litter fall. In the pine woodland this pathway is insignificant in comparison with the fluxes through the ground flora of bracken.

Temperate cultivated grassland

Temperate grasslands differ fundamentally from woodlands in the absence of woody biomass and the high biodegradability or food value of the leaf tissues for herbivores. The average total biomass for grasslands is about 1.6 kg m^{-2}, of which 50 per cent or more may be below ground according to grazing pressures, burning, etc. Net primary production is highly variable, ranging from 200 gm^{-2} in poor natural grasslands to over 1500 gm^{-2} in fertilized pastures.

The following account of the nutrient dynamics in a cultivated grassland is based on a review of this agro-ecosystem by Swift *et al.* (1979).

The simplified nitrogen flux pathways in a cattle grazed meadow are shown in Fig. 3.7. Figures are given for an unfertilized clover-grass sward and for a grass sward with added ammonium nitrate fertilizer. Both systems had a standing crop of nitrogen in the soil in the order of 1 kg m^{-2} which had accumulated over the years at a rate of about 5 gm^{-2} yr^{-1}. The fertilizer was applied at a level of 34 g N m^{-2} yr^{-1} and may be compared with a nitrogen fixation of about 16 gm^{-2} yr^{-1} by the clover root-symbionts. Nitrogen fixation by legumes can be in the order of 40 g N m^{-2} yr^{-1} but clover competes poorly with grass roots for a supply of phosphorus and tends to be eliminated from a sward in time if additional sources of phosphorus are not supplied. The fixed nitrogen is released from the clover to the soil by grazing and defaecation, exudation of organic nitrogen compounds from living roots, leaching from leaves and ultimately by the death and decomposition of the leaves. The grazing animals turned over about 10–25 g N m^{-2} yr^{-1}, of which 2–7 g was removed as dairy milk and 0.5–2 g N in fatstock. Most of the ingested nitrogen is therefore returned to the soil. The rumen of sheep and cattle is extremely efficient in breaking down the ingested leaf material and only about 10–20 per cent of the faeces is made up of undigested plant residues and over 50 per cent as bacterial cells. The faecal material decomposes rapidly and high rates of ammonification in relation to nitrification may result in the loss of nitrogen from the system as ammonia.

The total turnover of nitrogen in these grasslands is about 40 g m^2 yr^{-1} but this may represent only 1–4 per cent of the total soil nitrogen. As in the forest system, a significant fraction of the nitrogen is located in a

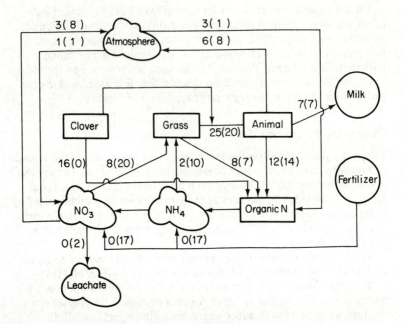

Fig. 3.7 Generalized flow diagram of transfers of nitrogen in gm^{-2} yr^{-1}, in cultivated temperate grassland grazed by dairy cattle. Transfers are for an unfertilized clover-grass sward and for a grass sward fertilized with about 34 g m^{-2} yr^{-1} of ammonium nitrate (the figures for the fertilized sward are in parentheses). The data are generalized figures from a wide literature and do not refer to specific sites. (Swift *et al.*, 1979; after Whitehead, D.C., 1970.)

slowly mobilized pool. The exchanges between the fast and slow cycling nitrogen pools are largely unquantified but tracer studies with the isotope ^{15}N have shown that up to 50 per cent of fertilizer applied to agricultural soils may not be recoverable in crops during the same growing season (Stewart *et al.*, 1963). Gaseous losses of fertilizer nitrogen through ammonification can also amount to 20–30 per cent under warm, moist conditions when oxygen tensions in the soil are low and nitrification is inhibited. Losses of fertilizer nitrogen as leachates are usually similar to natural ecosystems, less than 1 per cent of the annual flux, but higher losses may occur under heavy fertilization. This principally occurs in arable soils where the soil organic matter is low, the vegetation cover is not continuous and growth may be seasonally limited. Sandy soils, with a low capacity to absorb cations, are also likely to show higher leaching rates than soils with a moderate clay mineral content.

Attention has been directed towards nitrogen in this account but the

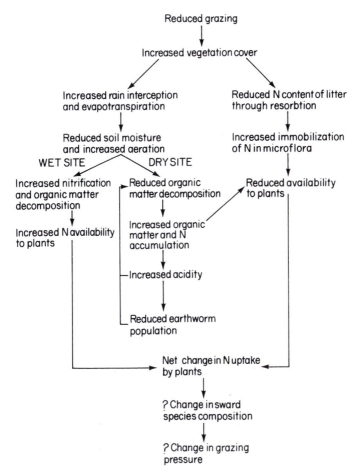

Fig. 3.8 Hypothetical sequence of changes in nitrogen transformations resulting from reduction in grazing on a pasture ecosystem. (From Swift *et al.*, 1979.)

importance of phosphorus for maintaining the clover biomass and hence symbiotic nitrogen fixation rate has been mentioned. The application of fertilizer with a high nitrogen content can also stimulate the growth of grass to a level where the clover is excluded from the pasture both by nutrient competition and shading. Changes in grazing pressures on pastures may have similar effects on nutrient cycling pathways and species composition (Fig. 3.8). The ecosystem is clearly a functionally integrated unit and changes in one major component is likely to bring about a reorganization into a new equilibrium point.

Aquatic ecosystems

Approximately 71 per cent of the earth's surface is covered by salt water
and only 2–3 per cent by freshwater bodies. Nonetheless, in this brief
account of aquatic ecology, discussion centres on freshwater systems for
the following reasons.
(1)   They are the most convenient and cheapest source of domestic
and industrial water.
(2)   Flowing waters are an important flux pathway for many biogeochemical processes.
(3)   Freshwater is a convenient and cheap means of waste disposal.
(4)   Lakes are significant sources of fish protein in many inland areas of
the world and are also the simplest aquatic systems to manipulate for fish
farming.
(5)   Running water is an important source of hydroelectric power in
many countries.
(6)   Lakes, rivers and streams have high amenity value.

Lakes
Different parts of lakes have different nutrient, oxygen, light and thermal
regimes and are inhabited by characteristic plant and animal communities.
The major life zones of a lake are shown in Fig. 3.9. The littoral zone
includes the shores (*eulittoral zone*) as well as the region below water
extending to the maximum depth of rooted vegetation (*infra-littoral
zone*). The term, littoral, is commonly used to refer to the infra-littoral
zone and will be used here in that context.
    The extent of the littoral zone is very variable in different types of lake
and is determined primarily by the lake profile. Steeply shelving lakes
have a narrow littoral zone and the contribution of macrophytes to NPP
is insignificant. In Lake Lawrence (Canada) however, a shallow lake where
over half the surface area is covered by macrophytes, equal amounts of
carbon are fixed by phytoplankton and the higher plants (Wetzel *et al.*,

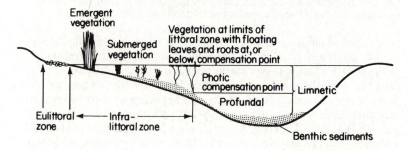

Fig. 3.9   Major life zones of a lake. See text for details.

1972). The development of plant and animal communities in the littoral zone is also influenced by the size of the lake and the degree of exposure of the shore. Wave action can create physical conditions similar to fast flowing streams. Exposed bays tend to have sandy shores and a pebbly substratum, poor in animal and plant species, while more sheltered bays in the same lake may have established aquatic vegetation, muddy bottom deposits and complex animal communities. A major proportion of macrophyte production is not grazed by herbivores, as in terrestrial systems, but passes on to a large and active decomposer subsystem. The majority of littoral animals are therefore detritivores or their predators.

In the open waters (*the limnetic region*) light intensity (I) decreases with depth according to the (approximate) function:

$$\frac{dI}{dt} = -kI$$

where k is an extinction constant ranging from 0.03 in pure water to 1.50 in muddy or eutrophic waters. Values of 0.20 to 0.40 are common in lakes and coastal waters. At some depth, according to local conditions, light intensity decreases to a point, the *compensation level*, where gross photosynthetic activity of plants is equal to their respiratory metabolism. The compensation level usually coincides with the lower limits of the littoral zone. Below this point (*the profundal zone*) no photosynthetic activity occurs and heterotrophs are supported by detritus from the sunlit photic zone (autochthonous detritus) or by terrestrial litter washed in by streams (allochthonous detritus). In the *photic zone* photosynthetic activity is carried on by phytoplankton (predominantly unicellular algae and diatoms) which supports a major herbivore subsystem (zooplankton and predatory fish). It has generally been assumed that production exceeds community respiration (P > R) in the photic zone and that dead materials then sink below the compensation level to the profundal zone where respiration exceeds production (P < R). While this may be a reasonable generalization, Golterman (1976) has shown that the decomposer subsystem can also be a significant functional component of the photic zone and 50–80 per cent of phytoplankton production may be mineralized by bacteria in surface waters.

The functioning of lake systems is extremely variable in space and time, particularly in the limnetic region where the seasonal availability of oxygen and nutrients is determined by the thermal properties of water. The maximum density of water is reached at 3.94°C and water at any other temperature will be layered on top unless external physical forces cause mixing.

Consider the situation in a deep temperate lake just after ice has melted in the spring and the water is uniformly at 4°C (Fig. 3.10). The oxygen concentration is approximately 13 ppm throughout the water column except for a small sag at depth caused by bacterial activity in the sediments

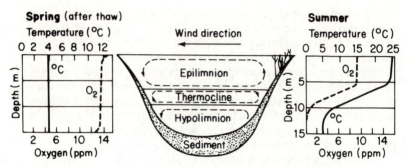

Fig. 3.10 Thermal stratification in a north temperate lake. For details see text.

and slightly higher oxygen levels at the surface due to enhanced absorption by ripples. As the seasons lengthen, heat is absorbed by the surface waters causing a rise in temperature but little heat is conducted to depth and the deep waters remain at approximately 4°C. In an absolutely still lake the temperature would decrease with depth down to the compensation level and then remain constant. Wind passing over the surface of the lake, however, causes mixing of the low density surface waters which circulate independently of the deeper, denser waters. When this occurs, the lake has become thermally stratified into an upper *epilimnion* and a lower *hypolimnion* separated by the *thermocline*. The thermocline and the compensation level do not necessarily coincide as a result of the surface water circulation. In a nutrient rich lake in the Netherlands, Vinkeveense Plassen, for example, the photic zone is only 2 m in depth while the epilimnion extends to 15 m as the surface is exposed to strong winds (Golterman, 1976). Contra-rotating water currents may be set up in the hypolimnion in response to the wind induced currents in the epilimnion but neither water, oxygen nor mineral nutrients are significantly exchanged across the thermocline. If the bottom sediments of the lake have a high biological oxygen demand, oxygen levels in the hypolimnion can decrease until anaerobic conditions are established and, over long periods of time, the bottom fauna may be excluded. This rarely occurs in temperate regions but some tropical lakes are more or less permanently stratified and the hypolimnion is azoic (without life) (Fig. 3.11). Under these conditions extensive denitrification to N_2 may occur and also the accumulation of reduced manganese and iron.

In temperate regions the surface waters cool in late summer, as the heat is lost from the water surface faster than it is absorbed, and finally reach a point where the epilimnion and hypolimnion equilibriate. The deep and surface waters of the lake can then mix or overturn under the influence of autumn winds. But as the temperature continues to decline to freezing point, a further thermocline is set up with the cold (below 4°C) surface waters of the epilimnion overlying a warmer hypolimnion. Equilibrium

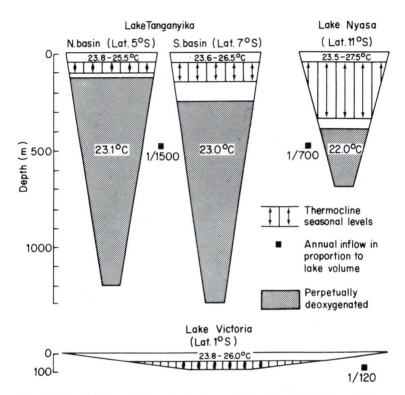

Fig. 3.11 Profiles (or vertical sections) of some African lakes. Note the extension and seasonal variation of the thermocline and the small differences in temperature between the lakes. (From Beauchamp, R. S. A. (1964). *Verh. int Verein. Theo. angew. Limnol.*, **15**, 91–9.)

and overturn occurs again in spring. Lakes with a warm summer and cold winter climate, which overturn in spring and autumn, are termed *dimictic*. Lakes with only one overturn a year are *monomictic*; this can occur in cold regions as the ice melts (cold monomictic lakes). Permanently stratified lakes are known as *meromictic*. Shallow lakes do not show thermal stratification even in tropical regions, for example, Lake Victoria (Fig. 3.11).

This overturn, characteristic of deeper lakes, has important consequences for primary and secondary production. In the littoral region nutrient exchange between the autotroph and heterotroph compartments is not spatially separated but in the limnetic region nutrients released by decomposition in the hypolimnion are not available to phytoplankton until overturn takes place. In dimictic lakes the phytoplankton shows an enormous burst of activity, or bloom, when turnover occurs. This rapid growth is checked as the nutrients in shortest supply become exhausted in the epilimnion. The commonest limiting nutrients are phosphorus, nitrogen and calcium, but other nutrients such as silicon can be important for particular groups or organisms, such as the diatoms which have a silicious test rather than a polysaccharide cell wall as in the algae. Many algae in the phytoplankton have adapted to this periodic shortage of some nutrients, notably phosphorus, by an excessive uptake of these elements in excess of their short term metabolic requirements. This *luxury uptake* is functionally similar to nutrient uptake by mycorrhizae and not only evens out the supply of limiting nutrients to the algae population but also reduces nutrient losses from the epilimnion during the growth season.

The zooplankton reach maximum numbers a month or so after the phytoplankton due to a delayed breeding response to excess food. By the time the zooplankton reach peak abundance the phytoplankton are beginning to deplete the nutrients in the epilimnion so that the rapid decline of the phytoplankton bloom is accentuated by the zooplankton grazing. This general scheme of linked seasonal zooplankton–phytoplankton oscillations is complicated by short term nutrient recycling mechanisms operating within the epilimnion. The grazing of phytoplankton mobilizes limiting nutrients, particularly phosphorus, through the excretory products of the zooplankton. Soluble phosphorus is also immobilized in bacterial cells and released by the feeding activities of protozoa. These internal recycling mechanisms are particularly important in meromictic lakes where nutrients in dead organic matter sinking below the thermocline are lost from the system. Similar nutrient mobilization pathways occur within the decomposer subsystem associated with lake and marine sediments (see reviews in Anderson and Macfadyen, 1976).

Lakes under natural conditions can be broadly classified into *oligotrophic* and *eutrophic* types according to their nutrient status and associated phenomena (Table 3.2). Typical oligotrophic (poorly nourished) lakes are clear and deep water bodies, usually found in regions of acid, igneous rock. The sides are often steeply shelving with a narrow littoral zone. Nutrient input from surrounding catchment area is low and production by phytoplankton is in the range $15-50$ g C m^{-2} yr^{-1}. Oligotrophic lakes become thermally stratified but phytoplankton do not show marked seasonal blooms because of the low availability of nitrogen and phosphorus. Phosphorus becomes bound by iron/humus colloidal complexes and is only released under anaerobic conditions when $Fe^{+++} \rightarrow Fe^{++}$. This is rarely achieved in oligotrophic lakes because the low organic matter content of the sediments has insufficient biological oxygen demand to deoxygenate the hypolimnion.

The eutrophic (well nourished) lakes are found in low-land regions and are comparatively shallow. Production by macrophytes and phytoplankton is high, at least ten times higher than oligotrophic lakes, due to a larger nutrient capital and input from the surrounding land. Thermal stratification may or may not occur according to depth but the deep organic

Table 3.2 Some characteristics of types of lakes. Typical ranges of characteristics for different types of lakes are given. Some lakes do not fit into this scheme, and additional types of lakes (sterile or ultra-oligotrophic, and polluted or hypertrophic but not saline) can be recognized. (From Whittaker, 1975.)

	Oligo-trophic	Meso-trophic	Eutrophic	Saline	Dys-trophic
Net primary productivity ($gm^{-2} yr^{-1}$)	15–50	50–150	150–500	500–2500	10–100
Phytoplankton biomass ($mg\ m^{-3}$)	20–200	200–600	600–10 000	1000–20 000	20–400
Total organic matter (ppm)	1–5	2–10	10–100	20–200	20–100
Chlorophyll a (ppb)	0.3–3	2–15	10–500	50–1000	0.01–2.0
Light penetration (m)	20–120	5–40	3–20	2–10	1–5
Total phosphorus (ppb)	<1–5	5–10	10–30	30–100	1–10
Inorganic nitrogen (ppb)	<1–200	200–400	300–650	400–5000	1–200
Total inorganic solutes (ppm)	2–20	10–200	100–500	1000–100 000	5–100

(ppm = mg 1^{-1}, ppb = μg 1^{-1} or m^{-3}; light penetration is the estimated depth to which one per cent of sunlight penetrates at midday)

bottom deposits usually become anaerobic and phosphorus is mobilized. Algae may be so prolific that the lake waters turn green, or even red according to the dominant species, at certain times of the year.

Oligotrophy and eutrophy are the extremes of a continuum of lake types. As a lake fills with sediment and the nutrient capital increases, it may develop from an oligotrophic to a eutrophic condition. There is some controversy as to whether this autogeny always takes place over a geological time scale since small lakes may undergo this transition in a few thousand years whereas Lake Baikal (U.S.S.R.) is still oligotrophic after 100 million years. These deep oligotrophic lakes have a high degree of global stability imparted by the nutrient sink in the bottom sediments. The nutrient enrichment of a lake may be accelerated by the influx of agricultural or domestic effluents (*cultural eutrophication*) but this development mimics rather than closely parallels natural eutrophication as a process of ecological succession. One of the best documented examples of cultural eutrophication is Lake Erie in the U.S.A. The Great Lakes may historically have been oligotrophic but Erie is comparatively shallow (about 66 m in depth compared with 200–400 m in the other lakes) and is fed by streams draining areas which are intensively cultivated or highly populated. Large quantities of nitrogen and phosphorus have been flushed into the lake and cause dense algal blooms in summer. During the day the epilimnion becomes super-saturated with oxygen but at night, when community respiration exceeds the O_2 tension, the surface waters may become deoxygenated and locally exclude the fish and invertebrate fauna. The hypolimnion is strongly anaerobic for most of the year. This maintains the nutrient mobilization from deep organic sediments formed by algal remains. The fresh water fisheries have been adversely affected by these conditions and the release of gaseous anaerobic decomposition products from the sediments, such as $H_2 S$, is an environmental nuisance to residential areas on the lake side. Strict legislation has reduced cultural eutrophication but the lake is probably too shallow and too polluted by heavy metals to recover its former equilibrium condition (c.f. Lake Washington, p. 5).

Rivers
Flowing water systems are characterized by a unidirectional current. This is a major variable determining the nature of the plant and animal communities and overrides other physical factors in the faster reaches. The relationship between current velocity and substrate characteristics of the stream bed is shown in Table 3.3. In addition to the flow velocity, the flow characteristics of water over the substratum are also important. Water flowing over a smooth even surface may show laminar flow in which the water molecules move parallel to one another without eddies or turbulence. This rarely occurs under natural conditions except in a thin layer, the boundary layer, immediately adjacent to the surface of a smooth rock. Away from the boundary layer the flow is turbulent and has a high erosive capacity depending upon current velocity. These flow characteristics have great significance for the distribution of aquatic animals and plants. Organisms sufficiently small or flat to occupy the boundary layer are protected from the full force of the current. A low pressure (wake) zone is also formed downstream of solid objects, which not only allows fish, such as trout, to maintain their position in the stream with minimum effort, but also sorts silt and organic particles from the stream flow forming an important microhabitat for aquatic animals. Small animals can also

Table 3.3 Size of objects moved by different current speeds. (From Nielsen, A. (1950). *Oikos,* **2**, 176−96.)

Speed of current (cm sec^{-1})	Diameter of objects moved (mm)	Classification of objects
10	0.2	Mud
25	1.3	Sand
50	5	Gravel
75	11	Coarse gravel
100	20	Pebbles
150	45	Small stones
200	80	Large stones (fist size)
300	180	Boulders

avoid the force of the current by living under the stones. In streams with near laminar flow the erosion forces are not only lower but less oxygen becomes dissolved in the water than at the same current speed with turbulent flow. These phenomena are of great significance in determining the patterns of settlement and biodegradation of organic pollutants.

Many schemes for temperate river zonation and classification have been proposed (see Hawkes, 1975). The following account is based on Carpenter's classification of upland streams. This scheme is not generally applicable to lotic (running water) systems but emphasizes the development of plant and animal communities in relation to stream flow characteristics.

Carpenter classified rivers into five zones – head streams, trout becks, minnow reaches, and the lowland courses of the upper and lower reaches. The ecological characteristics of the zones are outlined below:

(1) Head streams
Small volumes of water are usually present but these may reach torrent conditions in streams fed from glaciers or after snow melt. Head streams are oligotrophic, shallow, cool, fast flowing and saturated with oxygen. The substratum is bare rock or boulders and primary producers are encrusting algae or mosses, though macrophytes may be found in pools. The fauna is dominated by small animals, particularly Mayflies (Ephemeroptera) and Stoneflies (Plecoptera) living amongst the stones, epiphytes or deposits. There are few molluscs. The fish fauna is represented by small numbers of trout whose growth is often limited by food availability.

(2) Trout becks
These are formed by the union of head streams and have a more defined course and greater flow volume. The current is more uniform and the bed is of eroding rock. The waters are less oligotrophic but are still cold and saturated with oxygen. Vegetation is sparse and the animal community is primarily supported by allochthonous detritus. The strong current minimizes the development of plankton and limits the nekton (swimming animals) to strong swimmers such as trout and salmon. The benthic invertebrates are abundant and show obvious modifications to their mode of life. These modifications include dorso-ventral flattening to provide a large contact area with stones and to maintain the body in the boundary layer. Hooks and suckers are important attachment mechanisms in many insect groups. Net spinning Caddis larvae (Trichoptera) are important in this zone.

(3) Minnow reaches
As the current slackens erosion is reduced and deposition of coarse gravel may occur with deposits of coarse sand and detritus in pools and eddies. The water is usually saturated with oxygen but the temperature regime is higher and more variable. Filamentous algae, mosses and macrophytes are more abundant. A wide range of microhabitats and associated fauna is found in this zone. The benthic fauna is abundant, particularly the

Ephemeroptera (Mayflies), Plecoptera (Stoneflies) and Trichoptera (Caddis flies). Slower moving and free swimming insects such as beetles (Coleoptera) and bugs (Hemiptera) are found in slower reaches together with molluscs and crustacea. Burrowing insects and worms are found in the more silted regions. The characteristic fish species is the minnow (*Phoxinus*).

(4) and (5)   Lowland courses

The lowland courses are characterized by the deep river bed, large flow volumes, wide extremes of temperature and dissolved oxygen and the deposition of silt on the river bed. The fauna consists of many of the minnow-reach species together with larger numbers of molluscs, worms and leeches. Reduced oxygen levels may be found in the benthic sediments when temperatures are high and many benthic animals are physiologically adapted to these conditions. In slow flowing rivers the plankton may resemble that of small ponds. The fish fauna includes a number of slow swimming bottom living species.

The effects of pollutants on aquatic systems are usually manifest in terms of changes in the structure and functioning of the animal and plant communities. The use of biological indicators to monitor the combined and/or long term effects of pollution have therefore been used in water resource management studies since the early part of this century.

Local Water Authority and pollution biologists use a numerical classification of streams which enables them to present a summary of their findings to non-biologists alongside the corresponding physical and chemical data. Most biotic indices depend upon the general observation that in a polluted stream or river 'key' organisms, such as Plecoptera, Ephemeroptera and Trichoptera tend to disappear while chironomid fly larvae ('blood worms'), tubificid worms, some Crustacea and leeches increase in their relative abundance. Subdivision of these groups into families, genera or species increases the sensitivity of the diagnosis. The topic of biological indicators is comprehensively reviewed by Wihlm (1975).

The commonest form of water pollution is by organic matter such as sewage. This has the effect of stimulating bacteria and fungal growth (particularly the sewage fungus *Sphaerotilus*) by the provision of readily available carbon and nutrient sources. The enhanced microbial growth primarily results in a lowering of oxygen tension in the stream but the secondary effects of sediment and the flocculent growth of sewage fungus on stones and macrophytes also influence changes in the structure of the animal community. The *biological oxygen demand* (BOD) of a water sample is a measure of the amount of oxygen required for its biodegradation.

The effects of a sewage outfall on a lowland stream are illustrated in Fig. 3.12. The local effects of the effluent are moderately severe but once the concentrated effluent has been diluted and/or degraded by the pollution tolerant organisms, the river is eventually restored to its former condition above the outfall. Clearly the BOD of the effluent and any further effluent

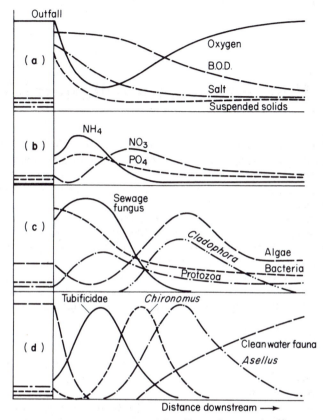

Fig 3.12 Diagrammatic representation of the effects of an organic effluent on a river and the changes as one passes downstream from the outfall: **(a)** and **(b)** physical and chemical changes; **(c)** changes in micro-organisms; **(d)** changes in larger animals. (From Mellanby, 1980.)

sources downstream will affect the length of river over which 'self cleansing' occurs, if at all. Lowland aquatic communities have a greater capacity to accommodate the perturbing effects of organic or thermal effluents than the oligotrophic, cold adapted communities of mountain streams. In most cases, however, recovery of the characteristic community for the reach is rapid following the cessation or reduction of effluent input. The recovery of a British river from gross organic pollution is described in detail by Brinkhurst (1965). A more comprehensive treatment of various aspects of river biology is given by Whitton (1975) and Townsend (1980). Hutchinson (1957) is a basic source of information on the structure and functioning of both lentic (still) and lotic (running) systems.

4 Populations and Communities

The functioning of an ecosystem is an expression of the interaction between the various plant and animal communities, or subsystems, which it contains. These communities can also be described in terms of their net functional attributes: primary production, the vegetation removed by grazing herbivores or the decomposition rate of organic materials. All of these effects are, in turn, the outcome of interactions within and between the populations forming these communities. It is therefore appropriate at this point to consider how the growth of plant and animal populations is determined by intrinsic factors and their physical environment, how they interact with other species, how these characteristics are integrated into the shaping of different communities and finally how these properties affect the functioning of ecosystems in space and time.

Delimiting Populations

A population may be defined as a group of organisms of the same species occupying a particular space at a particular time and may be described on worldwide and continental scales or at a local level. At each scale we can usually recognize groups of individuals with varying degrees of separation from one another: animals on islands, birds in patches of woodland, insects on different trees, plant species in clumps of vegetation. In some cases it is appropriate to consider the whole assemblage of groups as a population composed of subpopulations, or each subpopulation as a population, the particular definition depending on the objectives of the study. In any case the boundaries must be delimited before sampling can be carried out.

Sampling Populations

The population, once defined, may be sampled systematically to obtain absolute estimates of the density of individuals or the relative densities of two or more areas may be compared on a semi-quantitative basis.

It is rarely possible to take a complete census of individual plants or animals. Subsamples usually have to be taken, analysed and the results expressed on a unit area or volume basis. The size of the sample is determined not only by the density of individuals but also by their distribution. Plants and animals show distribution patterns which may be regular, random or clumped together according to the various influences of competitors, predators, microclimates, food, nutrients, breeding sites, etc. A series of samples might therefore all contain the same number of individuals, or numbers varying between zero and many individuals according to the degree of aggregation. In order to give an expression of the degree to which this sample variation affects the mean population estimate, it is conventional to take a large number of samples and present the mean counts with the upper and lower margins of error (fiducial limits). The number and size of samples required is usually a compromise between the accuracy of the estimate, the time and effort required for analysis and the damage which extensive sampling may do to the population and/or the habitat.

Populations are dynamic units and vary in time as well as space. The density of individuals in a population is a consequence of emigration and immigration rates as well as birth (natality) and death (mortality). Estimation of these parameters requires regular sampling at discrete intervals over a period which takes account of the life history of the species concerned.

Details of sampling rationale, methodology and statistical analysis are given by Southwood (1978) for animal populations and by Mueller-Dombois and Ellenberg (1974) for plant populations.

For some purposes it is sufficient to compare the relative densities of individuals in two or more areas. Semi-quantitative estimates can be obtained by the following methods.

(1) Traps, for example pitfall traps for ground living arthropods and sticky traps for flying insects. Migrating aphids can be continuously monitored by suction traps mounted on towers.
(2) Faecal pellets.
(3) Animal calls.
(4) Pelt return records such as those kept by the Hudson Bay Trading Company over 150 years.
(5) Artifacts, for example pupal cases, larval head capsules, bore holes of furniture bettles.
(6) Cover estimates of plant density.
(7) Frequency (presence or absence) of species occurrence in samples.
(8) Bait removal by animals, for example rodent pests.
(9) Catches per unit effort or time. This is a particularly appropriate method for aquatic surveys where sampling is difficult to carry out on an area or volume basis. Trawler returns per unit time or cost are also used as a measure of fish stocks and their commercial potential.

Age Structure and Mortality

Having determined the density of a population, what information can be obtained on its dynamic status from the samples of individuals? If the

animal can be sexed and aged, then the analysis is comparatively straight-
forward. Teeth, bones, claws, scales and otoliths (ear stones) in various
vertebrate groups show annual incremental rings like the growth rings in
trees. In very constant non-seasonal climates these characters may be
absent, as in tropical rain forest trees. In these situations, and also for
vertebrates and invertebrates which cannot be aged, weight or size classes
are carefully analysed to try and determine the groups or *cohorts* of individuals which resulted from a particular reproductive event. Similarly,
reproductive status can be readily determined in most vertebrates and
some invertebrates but in other cases it is impossible or may be estimated
from size or weight distributions.

Age structure pyramids for two human populations are shown in Fig.
4.1. These are the results of a single census and tentative conclusions
can be drawn about the dynamics of these populations. The pyramid for
Sweden has approximately parallel sides from the youngest to the post-
reproductive age groups. Thus it is reasonable to assume that the population is stable and when the parents die they are replaced by the same
number of offspring. In some societies where birth rates are low or large

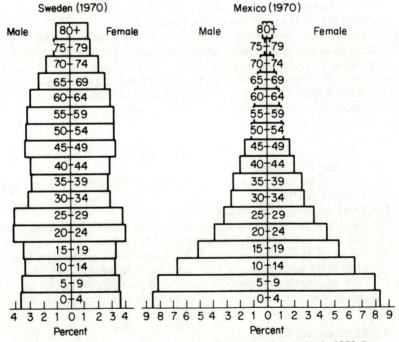

Fig. 4.1 Age structure of human populations in Sweden and Mexico in 1970. By
convention the males are plotted on the left and the females on the right. See text for
details. (From *U.N. Demographic Yearbook*, 1970/71.)

numbers of young people have emigrated, the age structure looks like the dome of a minaret. The constriction in the young age groups signals the decline and eventual extinction of these populations in the absence of immigration.

The population pyramid for Mexico has a broad base and a simplistic interpretation of its form is that when the cohorts of children move through to a reproductive age, the population will rapidly expand. Under conditions of low infant mortality and negligible birth control this would be true. However, in Malawi in 1971/72, for example, 64 per cent of deaths occurred before the age of five and a further 7 per cent boys and 14 per cent girls died between five and fourteen years. Life expectancy at birth was about 42 years (*U.N. Yearbook, 1977*). High birth rates compensate high rates of juvenile mortality and the population may therefore expand at nothing like its potential rate. Thus in order to describe the dynamics of a population, we must know not only the overall natality and mortality but when mortality occurs. In human populations, and some other long-lived vertebrates, this information can be summarized in the form of a life-table. *Instantaneous life-tables* can be constructed from the results of a single census. It is then assumed that the environment to which one cohort is exposed during the life of its members is the same as that for subsequent cohorts. If conditions of survival change markedly between cohorts (or if there is extensive immigration and/or emigration), then the instantaneous life-table may be invalidated. A *dynamic* life-table may be constructed for a single cohort (usually of 1000 individuals) whose history is documented from birth to death.

Both of these age specific methods of analysis have the same limitations in that future mortality and natality rates can only be estimated.

These data can also be summarized in the form of a survivorship curve which has the advantage, for the ecologist, that life histories of different organisms can be compared (Fig. 4.2). The first curve (Type I) is unusual in that most individuals survive and then die precipitantly. Starved flies in cultures approximate to this pattern. The second curve (Type II) is typical of human populations with high standards of medical care. Child mortality is low and adult mortality is spread over a number of years. In less fortunate human societies the survivorship curve becomes increasingly sigmoidal and approaches that of most other vertebrates which have a moderate to high juvenile mortality but good adult survival. In vertebrates survivorship curves range from this sigmoidal pattern to the extreme form of the third curve (Type III) where the majority of individuals in the population die while immature.

We will see that the general pattern of mortality experienced by a species has been accommodated by the evolution of appropriate reproductive rates (Fig. 4.5).

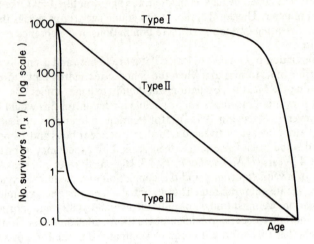

Fig. 4.2 Hypothetical survivorship curves. See text for details.

Population Growth and Reproduction

Population growth is an essentially multiplicative process and is continuous in situations where there is a complete overlap of generations. The growth rate of a population in the absence of restraint may be described by the simple logistic model:

$$\frac{dN}{dt} = rN$$

where N is the number of individuals in the population at an instant in time (t) and r is the maximum reproductive potential of an individual (the intrinsic rate of natural increase, sometimes designated r_m). If $r > 0$ there is exponential growth.

The rate of increase, r_m, is defined under optimal conditions of population density, food, space, light, temperature, humidity, and the absence of competitors and predators. Figure 4.3, for example, shows the variation in r_m for grain beetles under different temperatures and humidity conditions. This information is of considerable relevance for the storage of grain since excess food is potentially available to the insects under constant and favourable conditions.

All populations, however, exist in finite environments which ultimately limit growth rates. If the feedback on growth rate is proportional to population density, then the restraint is said to be density dependent. This can be expressed by a development of the logistic model:

$$\frac{dN}{dt} = rN\left(1 - \frac{N}{K}\right)$$

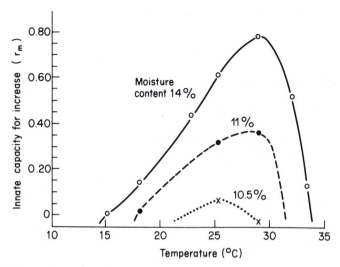

Fig. 4.3 Innate capacity for increase (r_m) of the grain beetle *Calandra oryzae* breeding in wheat of different moisture and temperature regimes. (After Birch, L.C. (1953). *Ecology*, **39**, 698–711.)

where K is the carrying capacity (the total number of individuals which can be supported under those conditions). A condition of this model is that the feedback mechanism is instantaneous and that there is complete overlap of generations. The exponential growth pattern is now modified by the additional function which generates a sigmoidal growth pattern and becomes asymptotic to K.

There are few occasions in which the provisors of this model are met. Two such situations, however, are the growth of yeast (Gause, 1932) and the floating pond-weed (Clatworthy and Harper, 1962) in cultures. In both cases the sigmoidal growth pattern is a close approximation to the logistic function. The feedback mechanisms are the production of alcohol, which limits the growth of the thin walled daughter yeast cells after budding, and mutual shading of the pond-weed fronds, which limits photosynthesis. Both mechanisms regulate the populations below levels where nutrients in the cultures are exhausted.

In most real situations the regulatory mechanism operates with a time lag caused by the time interval between successive generations, vegetation recovery after grazing and a host of other environmental phenomena. A time delay may be introduced into the logistic equation (Wangersky and Cunningham, 1956), which reflects the reaction time of the population to environmental change.

$$\frac{dN}{dt} = rN \left[K - \frac{N(t-T)}{K} \right]$$

where T is the time lag for the regulatory response. This function can be written

$$\frac{dN}{dt} = rNK - rN^2 \frac{(t - T)}{K}$$

Thus the regulatory component of the equation contains the expression N^2 which will produce a tendency to overshoot or under-compensate if the time delay in the feedback is long in proportion to the natural response time $(\frac{1}{r})$. This results in an oscillation about the equilibrium point, rather than a monotonic equilibrium with increasing numerical stability as the time lag increases. The population data for the grain beetles shown in Fig. 4.4 illustrates this type of oscillatory behaviour. It should be noted that the carrying capacity is set by the level of food resources available to these populations. Food is clearly limiting but the beetles do not reach a stable equilibrium as a result of delays in density dependent, regulatory mechanisms (egg production and predation, cannibalism of larvae and pupae by adults) which operate between generations. In populations which have higher values of r, shorter generation times and less complex regulatory mechanisms the cycles may be more extreme. Nicholson's (1957) blowfly populations show a good fit to this extreme pattern of cyclical 'boom and bust' oscillations.

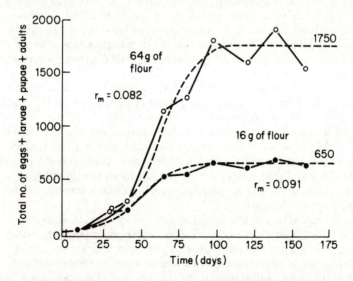

Fig. 4.4 Trends in the numbers of the flour beetle *Tribolium confusum* in 16 g and 64 g of flour at 27°C. The smooth curves are calculated using the logistic model. The observed points represent the combined numbers of all stages. (From Krebs (1972); after Gause, G. F. (1931). *American Naturalist*, **65**, 70–6.)

The logistic equations are an oversimplification of population growth and regulation in most laboratory populations, let alone natural populations, but serve to illustrate the basic principles of population dynamics. Detailed models pertaining to animal populations are reviewed in May (1976). Many of the principles of animal population dynamics have been applied to plant populations by Harper (1977).

The Regulation of Natural Populations

Plant and animal population densities are continually changing. Some species fluctuate irregularly over several orders of magnitude (insect pests and weeds), some show cyclical changes over relatively constant periods of time (small mammal populations or plant populations such as heather, see p. 98), while others show little variation in numbers from year to year (long lived species such as large vertebrates and trees). Nonetheless, for any particular habitat we are able to recognize a characteristic community of plant and animal species in which the rare species usually remain rare and the common species are generally common. Thus natural populations tend to fluctuate about a characteristic carrying capacity and are rarely observed to decline to extinction. This concept of the 'balance of nature' is one of the oldest tenets in natural history and can be traced back to the ancient Greek philosophers.

During the development of quantitative ecology in the early part of this century, it was recognized that the biotic and abiotic environment of natural populations determined the carrying capacity of a population and the degree to which it was regulated at this level. Views polarized, however, into two rather strongly opposed schools of thought.

The climatic school held that the density of animal populations, particularly insects, was regulated by density independent factors, primarily the weather. Central to this hypothesis is the concept that the growth rate of populations is limited by the shortage of time over which the rate of increase (r_m) is positive. Perturbations of temperature and moisture were considered to have a direct physiological feedback to reproduction and survival. Other species were regarded as having secondary importance to this effect.

The biotic school proposed that density dependent factors, particularly competition, regulated population size and that climate imposed secondary patterns of mortality on favourable breeding conditions. Climate was held to act as a density dependent factor in some cases, through the availability of refuges, but average population densities of species were not considered to be climatically determined.

The fact that this controversy raged over more than 50 years between a number of eminent ecologists reflects the complexity of animal population dynamics and the difficulty of identifying the operation of a single regulatory mechanism. Additionally, in cases where mechanism could be identified, there was a tendency to extend the principle uncritically to other populations which showed similar demographic patterns. The problem of differentiating between correlation and causation remains one of the most intractable problems in ecology.

Today the opposed viewpoints are largely reconciled and it is recognized that both density dependent regulation and climatic controls may predominate under particular environmental conditions. In environments which are physiologically favourable to a species, a constant or predictable climate, populations are primarily regulated by density dependent processes. At the other extreme, physiologically stressed environments, and/or climatic regimes which are unstable and unpredictable, show a strong influence of climatic conditions on population densities.

Closely integrated with this more comprehensive scheme is the concept of r and K selection (reviewed by Southwood, 1976). The r selected species have a high reproductive capacity and are typically associated with temporary habitats. Microorganisms, insect pests and weeds are typical representatives of this strategy where a rapid reproductive response is required to exploit temporary resources. Where the generation time and the period over which the habitat remains favourable is approximately equal, one generation will not affect the next and there will be no evolutionary penalty for an r strategist overshooting the carrying capacity (Southwood, 1976). Conversely K selected animals or plants occupying habitats where the environment is fairly constant will lower K and have an adverse effect on subsequent generations if the carrying capacity is substantially exceeded. A high value of r is a disadvantage under these conditions and selection operates for the quality, rather than the quantity, of individuals which are competitively superior. Some of the other correlates of r and K selection are shown in Table 4.1. One of the most significant of these is the relationship between r and generation time in organisms ranging from bacteria to man (Fig. 4.5).

Small organisms (e.g. bacteria) have a high 'r' which may be interpreted as an evolutionary adaptation to the multiplicity of factors imposing mortality on their populations. Larger organisms, such as vertebrates, have 'outgrown' some of these sources of mortality and their generation time extends over longer periods of time. This allows the possibility of parental care and protection as well as the long term survival of the species over periods when breeding conditions are unfavourable. It is also possible to see a range of r and K strategies within taxonomic groups which reflect adaptations to physically determined or biologically accommodated environments. Mean clutch size in samples of over 80 passerine bird species, for example, ranged from 2.3 in tropical forests to 5.6 in Central Europe (Southwood *et al.,* 1979). The larger clutch size reflects the greater variability of the European habitats and potential of species opportunistically exploiting temporal increases in food availability. The dandelion, *Taraxacum officinale*, similarly allocates a larger proportion

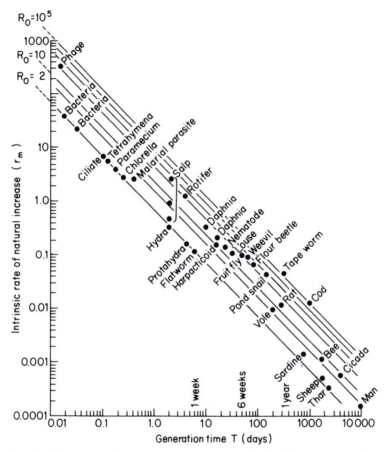

Fig. 4.5 The relationship between the intrinsic rate of natural increase (r_m) and generation time (T) in several organisms. The diagonal lines show various values of the net reproductive rate (R_0) where:

$$R_0 = \frac{\text{no. females born in generation } t + 1}{\text{no. females born in generation } t}$$

The rate of increase, r_m, is defined by the function

$$r_m = \frac{\log_e (R_0)}{T}$$

(From Heron, A. C. (1972). *Oecologia*, **10**, 269–93.)

Table 4.1 The contrasting suites of characteristics of the extremes of the r–K selection spectrum. (After Southwood, T. R. E. (1977). *Journal of Animal Ecology*, **46**, 337–66.)

r species	K species
Short generation time	Long generation time
Small size	Large size
High level of dispersal	Low level of dispersal
Much density independent mortality	High survival rate, especially of reproductive stages
High fecundity with low parental investment; often semelparous (breeding only once in a lifetime)	Low fecundity with high parental investment or iteroparous (breeding more than once in a lifetime) often with synchronous breeding or 'masting'
Panmictic	Territorial
Intraspecific competition often 'scramble' type	Intraspecific competition often 'contest' type
Low investment in 'defence' and other mechanisms of interspecific competition	High investment in 'defence' and other mechanisms of interspecific competition
Time efficient	Food and space efficient
Populations often 'overshoot' carrying capacity	Populations seldom 'overshoot' carrying capacity
Population density very variable	Population density more or less constant from generation to generation $\simeq$K
Habitat temporary, few generations in same location	Habitat permanent, many generations in same location

of biomass to reproduction in disturbed and transient habitats (Gadgil and Solbrig, 1972). Finally, different strategies may be adopted to the same gross environment within major taxonomic groups. The bony fishes (teleosts), for example, have massive egg production and high infant mortality in comparison with similar sized sharks and other elasmobranchs which give birth to a few, large, fully-developed young which have a high probability of survival. The continued co-existence of the two groups of fish in similar habitats clearly indicates that neither strategy is without advantages and disadvantages.

The Regulatory Processes

The complexity and interaction of the environmental variables influencing the density of an insect population are illustrated by the example shown in Fig. 4.6. It is likely that most r selected invertebrates show a similar complex of density-dependent and climatic influences upon their populations while K selected species are primarily regulated by density-dependent parameters. The following brief account of these processes is illustrated by examples where a single phenomenon can be seen to be operating. It should be borne in mind that this is rarely the case under natural conditions.

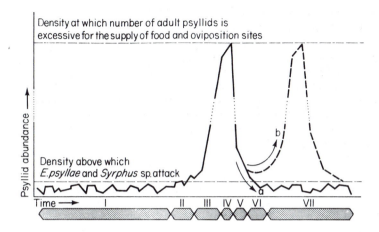

Fig. 4.6 Summary of the population changes in *Cardiaspina albitextura* in the Australian Capital Territory over an outbreak cycle. Stages within the cycle are indicated by roman numerals. Control mechanisms during each stage are as follows.
I Psyllid numbers either stabilized or restricted to slow increase by the combined action of (a) predation by birds, ants, encyrtid parasites, etc., on nymphs and eggs, the total percentage destroyed being independent of psyllid density; (b) the prevailing weather, especially temperature conditions; (c) density-dependent predation by birds on adult psyllids. **II** Stabilizing process fails, e.g. because unusually low temperatures reduce percentage parasitism. Psyllid numbers rise to a level at which the hyperparasite *Echtchroplexis psyllae* begins to destroy primary parasites. *Syrphus* sp. also attacks and destroys a few psyllid nymphs. **III** Psyllid numbers increase rapidly because of (a) decrease in percentage destruction by birds, ants, and encyrtid parasites; (b) failure of *Syrphys* sp. to compensate for waning influence of other predators. **IV** Environmental opposition to population growth increases again because of increasing damage to foliage by psyllid nymphs. **V** Damage to foliage very severe. Psyllid numbers decrease greatly because of (a) density-induced reduction in number of offspring per female; (b) number of offspring being excessive for the number of favourable feeding sites available on foliage. **VI** Psyllid numbers further reduced to an extent depending on (a) intensity of predation by birds, ants, encyrtid parasites, etc., which increases with decrease in psyllid numbers; (b) foliage replacement and associated shedding of infested leaves by the host plant. **VII** (a) If psyllid numbers are reduced below the level at which the hyperparasite *E. psyllae* operates, they are likely to remain low for some years; (b) if not reduced to this extent, numbers tend to increase rapidly to the level at which the available amount of foliage again becomes limiting. (Reprinted with permission from Clark, L.R. (1964). *Aust. J. Zoology,* **12**, 349–61.)

Intraspecific competition

Competition between animals of the same species can conveniently be divided into *scramble* and *contest* (Nicholson, 1957). Under conditions of scramble competition one individual is unlikely to gain the whole resource and under high population levels this form of interaction may result in insufficient of the resource for all. Ullyett's (1950) detailed work on blowfly populations is a good example of this phenomenon. Flies cultured with constant levels of meat showed changes in larval mortality (Fig. 4.7(a)), growth rates (Fig. 4.7(b)), size of the adult flies and fecundity (Fig. 4.7(c)) with increasing density and competition for food. These combined regulatory processes, however, were unable to check the population growth rate except at high densities. The fly populations therefore showed high amplitude cycles in which the few survivors of the population crash had superabundant food resources on which to breed. This cyclical phenomenon is an inherent property of populations which have inadequate or delayed homeostasis and occurs even under constant environmental conditions.

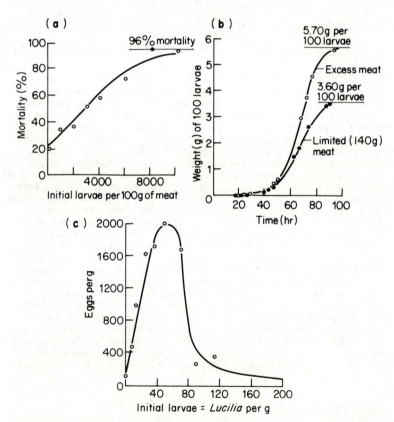

Fig. 4.7 Effects of intraspecific competition for food in blowfly (*Lucilia*) populations. **(a)** Total mortality as influenced by increasing larval density. **(b)** Effects of excess and limited food supply on larval growth. Note that the development times of the two groups of larvae are similar. **(c)** Numbers of eggs produced by adult flies after being subjected to different levels of larval competition for food. (Redrawn from Ullyett, G. C. (1950). *Phil. Trans. Roy. Soc., Lond. B.,* **234**, 77–174.)

In contest competition, the competitors interact in such a way that some individuals retreat or are eliminated so that the entire resource can be utilized by the victor. Contest is usually for breeding space, rather than for immediate food resources. In many vertebrate populations the contest may become ritualized so that the dispute can usually be settled without substantial damage to the contestants. A classical example of this form of interaction is a study of woodland owl populations by Southern (1970). The wood was divided up into a number of territories through intense vocal conflict between the owls. Only pairs of birds holding a territory were able to breed. The young birds were expelled from the woodland and formed a reserve population of animals which could only breed following the death or successful challenge of the resident pair. The owl population density within the wood remained constant and showed no synchrony with fluctuations in the abundance of their small animal prey. The survival of the young emigrant birds was, however, influenced by food availability. Similar examples of population regulation by the maintenance of breeding territories have been shown by Coulson (1968) for kittiwakes, Watson and Jenkins (1968) for red grouse and Carl (1971) for North American ground-squirrels.

Interspecific competition

Interspecific competition occurs between similar species which have similar ecological requirements and usually results, if the populations are resource limited and emigration is prevented, in the decline and ultimate extinction of one species. This phenomenon, often known as the competitive exclusion principle, has been extensively investigated in the laboratory with various invertebrates, particularly fruit flies and stored product pests. An example of competitive interaction between two species of flour beetle is shown in Fig. 4.8. Note that the outcome of the interaction is reversed by the presence of a parasite which affects one species more than the other. What is particularly interesting about some of these experiments is not so much that one species is eliminated from these highly artificial culture conditions, but that even under conditions of extreme crowding two species could co-exist if food, space or time could be partitioned in some way. For example, Crombie (1945) found that the grain beetles *Rhizopertha* and *Oryzaephilus* would coexist indefinitely because, although the adults had similar feeding habits, the larvae live and feed inside and outside the grains respectively. Barker (1971) showed that two fruit flies, *Drosophila melanogaster* and *D. simulans* can coexist in culture bottles under constant environmental conditions because space is used differently. *D. melanogaster* oviposits at the sides of the culture, the larvae live superficially in the medium and pupate the sides of the bottle. *D. simulans* oviposits in the middle of the culture, the larvae live at greater depth and pupate on the medium. Sometimes the resource division is

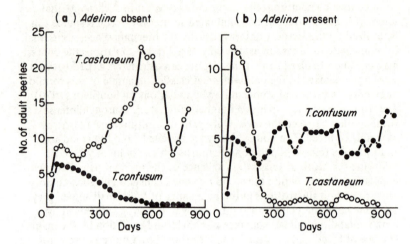

Fig. 4.8 Competition between two species of the flour beetle *Tribolium*. (a) The elimination of *T. confusum* in the absence of the parasite *Adelina;* (b) the reverse outcome when *Adelina* is present. *T. castaneum* was the more susceptible species. (Redrawn from Hassell (1976); after Park, T. (1948). *Ecological Monographs*, 18, 265–308.)

even more subtle. Merrel's (1951) experiments with *D. melanogaster* and *D. funebris* appeared to disprove the competitive exclusion principle but aging of the medium proved to favour first one species and then the other. At the end of the standard incubation period the two species were equally represented. A particularly important prerequisite of competitive exclusion, that environmental conditions are constant, was not therefore met in this experiment.

Under field conditions it is more appropriate to discuss competition between species in terms of their niche overlap. The niche has been variously defined as either the predominantly trophic role of an organism within a community (Elton, 1927) or the range of physical environments in which a species occurs (Grinnel, 1917). Hutchinson (1952) combined these concepts by defining the *fundamental niche* of a species as the extent of its total environment which it could potentially exploit in the absence of biotic restraints and the *realized niche* as the resources which it actually utilizes when other species are present. MacArthur (1968) suggested that in most cases interspecific competition could be described in terms of one or two parameters.

Under field conditions competitive exclusion is rarely observed except when species are introduced or colonize new habitats. The displacement of the red squirrel (*Sciurus vulgarus*) in the United Kingdom by the introduction of the North American grey squirrel (*S. carolinensis*) may be an example of this phenomenon, but forestry practice and the destruction of the deciduous woodland habitat of red squirrels is also involved. Alternatively the introduced species may be competitively inferior as is the case with many agricultural cultivars. Wild oats (*Avena fatna*) is a serious weed in Europe and North America where it competes with cereal crops and can cause a significant reduction in yield. Bell and Nalewaja (1968) have shown that wild oats persist in fields because its seed is shed before crop harvest and is not therefore removed by harvesting.

In most communities this type of interaction has occurred over an evolutionary time scale so that species with closely similar niches are found in separate habitats and food, or other resources, have been subdivided within habitats in such a way as to minimize competition. Lack (1944) found evidence of this niche segregation in most species of British passerine birds, Table 4.2, and considered that further study would reveal differences between the pairs with apparent overlap.

Table 4.2 Niche segregation of British passerine birds. (From Lack, 1944.)

	Cases (pairs)
Geographical separation	3
Separation by habitat	18 or more
Separated by feeding habits	4
Separated by size differences	5
Separated by different winter ranges	2
Apparent ecological overlap	5−7

Even with such established niche differences, competition is not a historical event but an ongoing dynamic phenomenon. This is rarely appreciated except when populations are experimentally manipulated. Grant (1969) investigated the ecological relationship between two North American vole species, *Microtus pensylvanicus* and *Clethrionomys gapperi*, characteristically found in grassland and shrubby woodland respectively. (The European *Microtus arvalis* and *Clethrionomys glareolus* show similar habitat relationships.) Experimental enclosures were set up containing equal areas of the two habitats. Using live traps *Microtus* was removed from the grassland area of some enclosures, and *Clethrionomys* from the woodland area of others. It was found that the *Clethrionomys* population moved out to occupy the grassland habitat more extensively than the movement of *Microtus* into the woodland. If the trapped animals were returned to their respective habitats then the original distribution of the two species was re-established. The interaction occurred over a short period of time indicating that it was a behavioural interference mechanism operating, rather than resource exploitation. The realized niches of both species are therefore smaller than their fundamental niches but that of *Microtus* (the grassland specialist) is smaller and included within that of

Clethrionomys (the generalist). Thus there is evidence of a competitive interaction between these species notwithstanding the nich differences. However, intraspecific competition, predation, disease and climate regulate both populations well below levels where competitive displacement is likely to occur.

It is important to recognize that there are conditions (listed below) where closely similar species can co-exist in the same habitat without competitive displacement taking place.

(1) In environments which never establish equilibrium conditions which favour a single species (this includes temporary habitats and ephemeral (short-lived) resources such as carrion and decomposing plant materials).

(2) Fluctuating environments which reverse the competitive balance before extinction can occur. There are many examples of this phenomenon in plant communities where different species are favoured by wet or dry conditions.

(3) Environments with extreme and variable climates which limit population growth rates.

(4) Conditions where the populations of potential competitors are strongly regulated by predators or parasites (see Fig. 4.8).

Predation

It would appear reasonable that predators regulate their prey populations and that prey availability regulates the predators. In fact, this is difficult to demonstrate in natural populations because of the influence of other environmental variables such as competition, disease and climate on both predator and prey populations. Classical examples of apparently reciprocal density dependent interactions, such as the snowshoe hare—lynx population cycles (Fig. 4.9) have proved, on re-examination, that while the

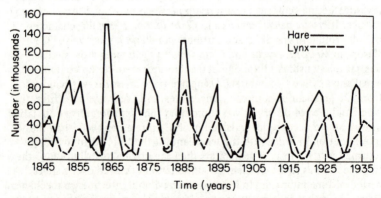

Fig. 4.9 Cyclical oscillations in the abundance of lynx and snowshoe hares based on pelt records from the Hudson Bay Company of Canada. (After MacLulich, D. A. (1937). University of Toronto Studies, Biological Series No. 43, pp. 1–136.)

predator is largely dependent upon the prey for food it is not regulating the prey population. Indeed, similar oscillations with a 10-year periodicity occur in hare populations on islands in northern Canada in the absence of predation. Cyclical oscillations of small mammal populations are discussed in detail by Krebs and Myers (1974); Caughley (1970) considers similar phenomena in wild ungulate populations. Most predators are also polyphagous (i.e. they do not exhibit prey specificity) and this further complicates the analysis of predator/prey interactions under field conditions.

Much of the research on predation has been carried out on invertebrate predators and particularly parasitoids (insect parasites which kill the host) which have simple life histories and are amenable to laboratory study. Additional impetus to this line of research is given by the search for effective biological control measures for agricultural pests as an alternative to the use of pesticides, such as the example shown in Fig. 4.10. Biological control can also involve the use of herbivorous insects for weed control,

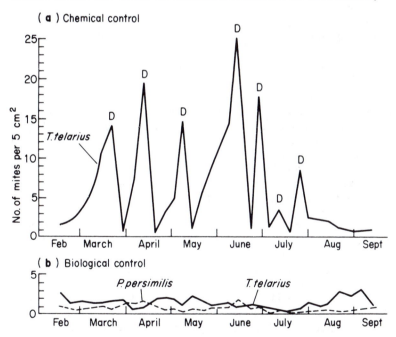

Fig. 4.10 The effect of **(a)** chemical and **(b)** biological control on the spider mite, *Tetranychus telarius*, in a commercial cucumber crop in 1967. Difocol treatments (marked D) were used when *T. telarius* increased to a level causing damage. *Phytoseiulus persimilis*, a predatory mite, was placed in another part of the greenhouse when damage to cucumber plants was first noticed, and transferred from one plant to another at weekly intervals. The biological control was not only more effective in a larger number of outbreaks but the cost was about 74 per cent that of the chemical method. (From Markkula, M. *et al.* (1972). *Annales Agriculturae Fenniae*, 11, 74–8.)

bacteria and viral pathogens against insect pests and a host of insect parasitoids.

An ideal biological control agent will rapidly reduce the population of the target organism and then remain in stable, low density equilibrium preventing further outbreaks. This is illustrated by the control of the prickly pear cactus, *Opuntia* spp., in Australia by the moth *Cactoblastis cactorum*. After its introduction as a hedging plant in 1839, *Opuntia* spread rapidly: a single root-stock can increase to about 625 t h^{-1} in two years (Dodd, 1940) and by 1925 the cactus had taken over some 12 million hectares of land. Extensive stands of the cactus, at an average density of 12 500 plants ha^{-1}, seriously reduced sheep and cattle grazing. *Cactoblastis* was introduced from South America in 1925 and within two years *Opuntia* was virtually eliminated and a tight grazing equilibrium was established at a low level of about 30 plants ha^{-1} (Dodd, 1940). Mathematical aspects of this and other plant-herbivore interactions are reviewed by Caughley (1976).

Laboratory predation studies have revealed a large number of parameters which affect the stability of predator/prey interactions. These include predator's prey specificity, the overlap and dynamics of the predator and prey life cycles, the importance of refuges where the prey can escape predation and interference between predators.

Two important determinants of the characteristics of predator/prey interactions are the numerical and functional responses of the predator. The numerical response is the change in predator density by reproduction or immigration while the functional response is the influence of prey density on the rate of prey capture.

The basic functional response curve varies in form according to the behaviour of the predator (Fig. 4.11) but in each case there is a level of prey density beyond which the predator cannot increase the number of prey attacked.

The 'Type I' response (Fig. 4.11(a)), shown by filter feeding animals, depends upon a constant rate of prey (food particle) encounter which reaches a threshold at the maximum feeding rate of the animal. The 'Type II' response (Fig. 4.11(b)) is widespread among insect parasitoids and predators. The curve of the response is imparted by the handling time (time taken to catch, kill and eat the prey) which is a constant and reduces the time available for further search.

The numbers of prey attacked (N_A) is given by the function:

$$N_A = \frac{a'NT}{1 + a'T_HN}$$

where a' is the attack coefficient (a constant rate of prey encounter), T is the whole time available, T_H the handling time and N is the total prey population. This is known as the *disc equation* because Holling (1959) explored its characteristics using blindfold human subjects (predators)

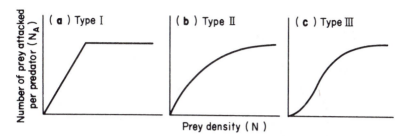

Fig. 4.11 Three types of functional response showing the changes in the number of prey attacked per unit time by a single predator as the initial prey density is varied. See text for details. (Redrawn from Hassell (1976); after Holling, C. G. (1959). *Canadian Entomologist, 91*, 293–320.)

searching for sand paper discs (prey). The functions a′ and T_H vary between the developmental stages of the predator and with prey size.

The 'Type III' sigmoidal response reflects a learning ability on the part of the predator, illustrated by the experiments of Holling (1965) where a deermouse, searching for sawfly pupae (the preferred food) buried in sand at various densities, was provided with an excess of readily available, but less acceptable, food in the form of dog biscuits. As the sawfly density increased the mouse learnt to search for the preferred food. This type of behaviour is common in birds — when prey density is low they will concentrate on other prey or move to alternative areas where prey density may be higher (Royama, 1970).

For parasitoids the numerical (breeding) response is a linear function of hosts parasitized, but for most other predators it shows a similar pattern to the 'Type II' response curve.

The combination of the two responses determines the basic characteristics of the predator's capacity to respond to increases in prey density. If the fecundity and generation time of the predator and prey are substantially different, as in an r-selected invertebrate and a comparatively K-selected vertebrate predator, then predation may not exert a significant influence over the prey population except at low endemic densities (Fig. 4.12). Closely similar life histories can also result in an unstable equilibrium. Huffaker (1958) showed that if a predatory mite, *Typhlodromus occidentalis*, was added to a population of the suctorial mite, *Eotetranychus sexmaculatus*, feeding on a single orange, a single oscillation occurred and both populations became extinct. The experimental system was then increased to 250 oranges connected in various ways to form a three-dimensional habitat complex in which the prey could breed and emigrate to other parts of the complex before the predators could locate and decimate local aggregations. The result was a series of stable oscillations over a period of 70 weeks before the predator died out. In this case the spatial 'prey refuge' ultimately defeated the predator but it is clear that

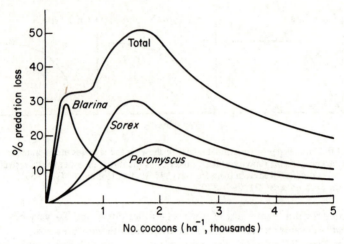

Fig. 4.12 Combined functional and numerical responses for small mammals predating cocoons of the European pine sawfly. (After Holling, C. S. (1959). *Canadian Entomologist*, **91**, 293–320.)

the spatial complexity conferred an element of stability on what was otherwise an unstable relationship.

Climate and weather

Weather extremes can cause mortality in most animal and plant populations. Lack (1966) cites an example of two heron (*Ardea cinerea*) populations, in widely separated regions of Great Britain, which showed similar changes in abundance caused by the exceptionally hard winters of the 1940s and 1960s. Normally, homeotherms (warm-blooded animals) are little influenced by weather, though changes in food abundance and quality may have indirect effects on their populations. Invertebrate populations, however, are strongly influenced by the effects of temperature, moisture and daylength on their physiology and behaviour, as well as mortality induced by unfavourable conditions (see Fig. 4.6). These aspects of insect population ecology are considered in detail by Varley *et al.*, (1973).

Weather can also have important indirect effects on insect populations by increasing their susceptibility to disease. The diamond-back moth, *Plutella maculipennis*, is a widespread pest of cabbage and other Brassica crops. Ullyett (1947) showed that, in South Africa, the onset of the rains produced conditions suitable for the rapid transmission of a fungal pathogen which destroy the caterpillar population. The wet conditions also affected the parasitoid populations and the percentage of parasitized larvae declined over this period. After the rains the number of caterpillars per cabbage increased rapidly before the parasitoid population could respond to the increase in host abundance.

Food quality

The amount of food available to an animal population is of paramount importance. Food availability affects territoriality and the growth and survival of young, and competition for food leads to the ecological separation of species. Over the last few decades, however, ecologists have come to realize that the quality of food, not just its quantity, has an important influence on animal populations, particularly in determining the carrying capacity of a habitat.

All organisms require the same basic range of elements for growth and maintenance, but while plants can use inorganic sources of minerals, herbivores are largely dependent on plants for nutrients, predators upon other animals, and decomposers upon dead animal and plant remains. The importance of nutrient availability to plants has long been recognized but variations in the nutrient content of the plants themselves influence the dependent herbivore and decomposer subsystems. This effect is most noticeable at the plant/herbivore interface because while the tissues of a starving herbivore may be suitable food for a predator, a nutrient deficient plant may have adverse effects on the health of a herbivore. This is because of the far greater dissimilarity between the composition of plant and animal tissues than between different animal tissues.

House (1961, 1966) defines food as having three major attributes: physical feeding requirements, chemical feeding requirements and the nutritional requirement of the animal. These principles were described for insect populations but are discussed here in a more general context.

The physical feeding requirements include texture and morphology of the food plant. Bernays and Chapman (1970) showed that the nymphal instars of the grasshopper *Chorthippus parallelus* were affected in different ways by the grasses *Festuca* and *Holcus*. The nymphs were unable to feed on *Festuca* because of the thickness of the leaf blades and unable to feed on *Holcus* because of the dense covering of silicious hairs (trichomes) on the leaf edge. Trichomes are effective defense against many herbivorous insects (Levin, 1969) and even snails, attracted to plants by their odour, may be prevented from climbing the stem (Grime *et al.*, 1970).

The chemical attributes of plants include a range of secondary compounds which are not nutritional requirements for insects but are produced as a defence against attack by herbivores. The nature and effects of these compounds have been comprehensively reviewed by Whittaker and Feeny (1971). Some herbivorous insects have adapted to the presence of these compounds by synchronizing their life cycles to the growth cycle of the plant. The oaks (*Quercus* spp.), like most other trees, have high concentrations of tannins in their mature foliage which precipitate proteins, including the digestive enzymes of the larval winter moth (*Opheroptera brumata*). The larvae develop rapidly on the leaves as they burst from the buds, when the tannin load is low, and pupate at the time when the leaves reach maturity. Other insects detoxify or tolerate the toxin and even isolate it in their bodies as a basis for their own chemical defenses. This specialization of an insect species imposes a dependence on a particular food plant. The chrysomelid beetle *Chrysolina*, for example, feeds on the St. John's wort *Hypericum*, which contains the toxic alkaloid hypericin. The beetles are not only unable to feed in the absence of hypericin but will attempt to feed on anything to which this phagostimulant has been added. *Chrysolina* have been spectacularly successful in America for the biological control of *H. perforatum* which is poisonous to stock. These and other aspects of the close evolutionary relationships which have co-evolved between animals and plants are covered by Gilbert and Raven (1975).

Most of the examples of the nutritional requirements of animals affecting their populations concern nitrogen (or amino acids) but any one of the mineral macronutrients or trace elements (see Table 2.2) can be involved (see Underwood, 1971 for a comprehensive treatment of trace elements in human and stock nutrition). Seasonal changes of aphid populations, for example, are related to the amino acid content of the plant sap (Dixon, 1970). Rapid growth of the aphids occurs in spring when nutrients are being translocated to the leaves for growth, and in autumn when withdrawal from the leaves occurs before leaf fall. In the summer months the aphid populations are nutrient limited and have a by-pass system in the gut which allows the rapid throughput of sap to increase nitrogen uptake. The excess sugars are secreted as 'honey dew'.

The consequences of diet quality and/or selective feeding may not be confined within the interaction between the autotroph and herbivore subsystems but transferred to other components of the animal community. The Australian bushfly (*Musca vetutissima*) breeds in cowdung and seasonal changes in quality are reflected by the responses in the insect population (sex ratios, size, fecundity, survival and dispersal) shown in Fig. 4.13. Similar effects of resource quality can be identified elsewhere in the decomposer subsystem, such as the growth of bacteria and fungi, and the growth response of the animals which feed on them. Ultimately this results in the release of nutrients back to the plant subsystem after a time delay. This lagged nutrient cycling response has been proposed for the regulatory mechanism underlying the lemming cycles in tundra ecosystems but this hypothesis, among others, has been dismissed by Krebs and Myers (1974) for lack of evidence.

Human Populations

There are no accurate records of human populations before the first censuses were carried out in the 17th Century but since then the population

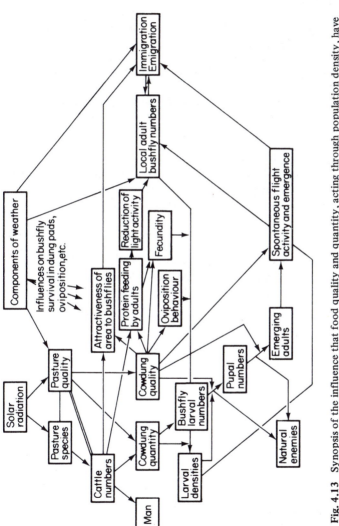

Fig. 4.13 Synopsis of the influence that food quality and quantity, acting through population density, have on the life system of the Australian bushfly. Dung pad deposition rates are more or less constant throughout the year but resource quality shows marked seasonal variation. Food quality not only affects the survival and reproduction of the adults but also their flight activity: the flies remaining longer in areas with high quality food and dispersing from areas where the food quality is low. (From Hughes and Walker, 1970.)

has doubled and redoubled at an ever increasing rate which is actually faster than exponential growth (Fig. 4.14). The world population in 1650 was about 500 million and it doubled over the next 200 years. It took only 80 years from 1850 to 1930 to double again and 45 years to reach a world population of about 4 billion people in 1975. Projections of world population growth are that the next doubling will take 25–36 years, at present rates of increase of about 2 per cent per annum, and that 6 to 7 billion people will be on earth at the turn of the century. What happens beyond that time is highly speculative but there is little doubt that the carrying capacity of this planet will be reached or exceeded within the lifetimes of children currently being born.

The major difficulty in forecasting world population trends is the differences in the growth rate of populations in different parts of the world and changes in birth and death rates which can occur over quite short periods of time.

Demographers estimate population changes from separate calculations of births and deaths per 1000 individuals and the difference, the growth rate or rate of natural increase 'r', is then expressed as a percentage. Human populations up to 1975 in different parts of the world and per-centage growth rates are shown in Table 4.3. The more developed areas of the world have a growth rate of less than 1 per cent, equivalent to a doubl-ing time of 100 years or more, while less developed countries have growth rates of more than 2.5 per cent and will double in less than 30 years.

The reasons for both the rapid increase in world population and the lower growth rates in the developed countries are complex and described

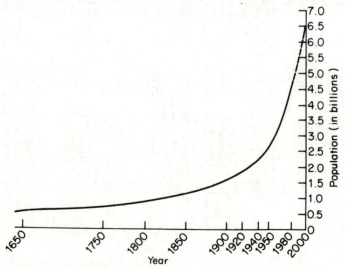

Fig. 4.14 World population growth since the 17th Century.

Table 4.3 World populations and annual growth rates. (From Ehrlich *et al.*, 1977.)

| | Population (millions) | | | Annual growth rate (%) | |
	1960	1970	1975	1960–1970	1970–1975
WORLD TOTAL	2995	3621	3988	1.90	1.93
More developed regions*	976	1084	1133	1.05	0.88
Less developed regions	2019	2537	2855	2.28	2.36
EUROPE*	425	459	474	0.77	0.64
Eastern Europe*	97	103	106	0.62	0.64
Northern Europe*	76	80	82	0.57	0.41
Southern Europe*	118	128	132	0.78	0.74
Western Europe*	135	148	153	0.96	0.67
U.S.S.R.*	214	243	255	1.25	0.99
U.S.A. AND CANADA*	199	226	237	1.31	0.90
OCEANIA	16	19	21	2.12	1.98
Australia and New Zealand*	13	15	17	1.92	1.83
Melanesia	2	3	3	2.44	2.56
Micronesia and Polynesia	1	1	1	3.42	2.64
SOUTH ASIA	865	1111	1268	2.50	2.64
Eastern South Asia	219	285	326	2.65	2.69
Middle South Asia	588	749	853	2.41	2.59
Western South Asia	58	77	90	2.79	2.96
EAST ASIA	787	926	1005	1.62	1.63
China	654	772	838	1.70	1.64
Japan*	94	104	111	1.03	1.26
Other East Asia	39	50	56	2.39	2.15
AFRICA	272	352	402	2.58	2.66
Eastern Africa	77	100	114	2.60	2.71
Middle Africa	32	40	45	2.41	2.35
Northern Africa	65	86	99	2.77	2.81
Southern Africa	18	24	28	2.87	2.70
Western Africa	80	101	116	2.36	2.59
LATIN AMERICA	216	284	326	2.74	2.73
Caribbean	21	26	28	2.12	2.14
Middle America	49	67	79	3.19	3.21
Temperate South America*	31	36	39	1.64	1.44
Tropical South America	116	155	180	2.93	2.91

*The regions marked with asterisks are considered as 'more developed' from a demographic point of view.

in detail by Ehrlich *et al.* (1977). It is generally considered that the populations of primitive man were held in check by disease, warfare and the limited and uncertain food returns of a hunter–gatherer existence. Following the agricultural revolution 7–9 thousand years ago human populations became settled and started producing an excess of food resources over their immediate needs. Food could be stored against hard times, the standard of nutrition improved and death rates declined. This pattern was accelerated by the industrial revolution which further differentiated populations into rural food producers and urban consumers. Agriculture became energy subsidized instead of labour intensive. Fertilizers improved both yield and quality of crops. Human welfare was improved by education, medicines, higher standards of hygiene and nutrition. The life expectancy at birth has increased by 40 per cent between 1950 and 1975 in the developing world, by a further 12 per cent in the developed world and by 35 per cent in the world as a whole. Thus the gains in living standards made in the United Kingdom and other industrialized countries during the 19th and 20th Centuries are now being made in developing countries (Allaby, 1977). Against this pattern of declining death rates the birth rates in most developing countries have remained comparatively constant and the populations have grown. The age structure develops a distinct pyramidal structure (see Fig. 4.1) which points to continued rapid population growth as the juvenile cohorts move through to reproductive age. In Mexico 46 per cent of the female population is below 15 years of age, the population growth is 3.4 per cent per annum and will double in 20 years. In the majority of developed countries, however, birth rates have declined. European countries 150 years ago had high birth and death rates but at the turn of the 19th Century the annual death rate fell from about 40/1000 to 10/1000. Some 50 years later the birth rate declined from 40–50/1000 to around 13/1000. The reason for this is a change in human fertility rather than fecundity. Fecundity is the child bearing potential of a woman: a rate of one birth per 9–11 months over the child-bearing years of 15 to 50. Fertility is the actual number of children born. In the developed countries the trend towards later marriages (associated with longer periods of education and/or employment of women) and smaller numbers of children (facilitated by contraception) has stabilized the age structure and population growth.

There are some signs that the rate of world population growth has slackened since 1970. This is attributed to (1) the unexpectedly successful results of China's population policies; (2) a significant decline in the birth rates of almost all developed countries; (3) some success in lowering birth rates in some less developed countries; and (4) a rise in death rates in several countries, particularly in South Asia, caused primarily by food shortages during the famines of the 1970's (L.R. Brown in Ehrlich *et al.*, 1977).

The common factors in the less developed countries which are undergoing a demographic transition (including Taiwan, Egypt, Chile, South Korea and Cuba) seem to be that the majority of people have experienced an increase in social welfare (literacy, health care, reduced infant mortality and better diets) in spite of the fact that these countries are very poor (Murdoch, 1975). If this encouraging trend spread to other developing countries the world could stabilize around 6 billion but the extent to which this reduced growth is caused by a rise in death rates is a tragic cause for continued concern over world population growth in relation to food resources.

Characteristics of Communities

The term community is used in the ecological literature to refer to many different scales of species assemblages ranging from small groups of animals or plants to the total biota or species complement of organisms in an ecosystem. These species assemblages often merge continuously into one another along gradients, so that it may be impossible to tell where one community ends and the next begins unless there are physical discontinuities between habitats. These discontinuities may be the limits of a moss cushion or a rotting log, a change in soil type or parent rock, some local topographical feature or, for the gross ecological community, the physical limits between one ecosystem type and another. At the intersection between two dissimilar ecosystem types there is usually a transitional zone, the *ecotone*, which is a region of high biological activity, resulting from the interaction of the two ecosystem biotas as well as groups of ecotone specialists. This 'edge effect' is characteristic of the littoral zone between terrestrial and aquatic systems and of the shrubland separating grassland from forests. Hedges are important ecotones in agroecosystems; they not only form refuges for wildlife in general but also a reservoir of predators and parasites of agricultural pests — as well as many of pest species themselves.

Whatever the scale of the community the integrated relationships between the component species is implicit in the concept: a community has attributes which do not reside in the individual species. These characteristics of animal and plant communities may be summarized as follows.

1. Growth form and structure of the plant community The community can be described by the major categories of plant growth forms (trees, shrubs, herbs, mosses and algae) which constitute the physical structure of the plant community.

2. Species diversity This is the number of species in a community (species richness) and their relative abundance or equitability. The dominant species may exert a controlling influence over the community by virtue of their size, numbers or activities.

3. Trophic structure The feeding relationships within and between the subsystems.

Growth form and physical structure of plant communities

Terrestrial plant communities vary in structural complexity from the rock encrusting and cushion-forming growth forms of arctic-alpine environments to the vertical layering of forests into the canopy, sub-canopy, shrub and herb layers. Associated with this increase in vertical stratification is an increase in primary productivity (photosynthetic tissues show greater differentiation into sun and shade adapted forms in forests) an increase in microhabitat complexity (providing more habitable space for epiphytes and animal life) and a range of buffered microclimates (Fig. 4.15) which are important for animal and plant species.

In aquatic habitats water provides the support for phytoplankton and they do not contribute to the structural complexity of the habitat. Coral reefs and submerged aquatic macrophytes provide a habitat complex for animals but buffering of climatic extremes is, of course, conferred by the surrounding water and not by the vegetation complex.

Species diversity

The number of species in a habitat is a simple measure of its diversity but the distribution of numbers of biomass between the species is more ecologically meaningful. The term *species diversity* is usually accepted to mean both species richness and the distribution of numbers or biomass between species (equitability). Two extreme forms of distribution are observed in natural habitats: a low number of species with high dominance

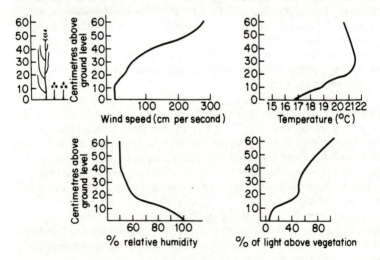

Fig. 4.15 Diagram showing the structure of grassland vegetation (top left) and the effect which this has upon the microclimate of the habitat. (From Cox, C. B. *et al.* (1973). *Biogeography*. Blackwells, Oxford.)

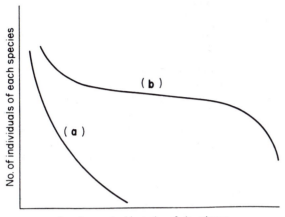

Fig. 4.16 Two extreme conditions of species diversity in communities. (a) Low number of species with dominance by one or two species (low equitability). This type of distribution is characteristic of communities in extreme environments. (b) High species richness and equitability characteristic of favourable and constant environments.

of one or two species, and a large number of species with a high degree of equitability (Fig. 4.16). These species distribution curves are approximated by the negative binomial and log normal distributions respectively with a range of other patterns of distribution in between. Various numerical indices have been proposed to quantify species diversity and most suffer from one defect or another – a lack of sensitivity to rare or common species, or a greater dependence on one type of distribution than another. There is a vast ecological literature, devoted to the evaluation of the relative merits of a large number of species diversity indices, which has been reviewed by Pielou (1975). There is always a loss of biological information when a species distribution curve is expressed as a single index and initial evaluation of the species diversity characteristics of a community is best carried out graphically.

One of the most interesting and complex questions to which ecologists have been directing their attention over the years is why there are more species in one habitat than another. In particular, why gradients in the number of species, and equitability, can be observed from the poles to the equator. Arctic habitats are characterized by having few species with a high degree of dominance while tropical rain forests are highly species rich and equitable. This is a complex topic with a large supporting literature. General accounts will be found in Krebs (1978) and Whittaker (1975). A large number of hypotheses have been proposed to account for latitudinal gradients in species diversity, or differences in species diversity between habitats, and these can be summarized under six major groups of phenomena: the time hypothesis, competition, productivity, predation, spatial heterogeneity and environmental stability (Pianka, 1967). Few of the hypotheses have been tested experimentally.

1. The time hypothesis   All communities increase in diversity with time, either through the evolution of species or on a shorter time scale by the immigration of species into suitable habitats. Speciation rates are assumed to be faster in the favourable conditions of the humid tropics and hence diverse biotas have developed. Slower speciation rates in temperate and polar regions, interrupted by glaciation and climatic disasters, have resulted in the formation of relatively immature communities (low diversity and equitability).

2. Competition   Communities in regions of extreme climatic conditions are assumed to be mainly shaped by physical environmental parameters rather than by biotic interactions. Under more favourable conditions competition is more intense, species are more specialized and more niches can be defined within a given habitat.

3. Productivity   The basic hypothesis states that greater production results in greater community diversity.

There is circumstantial evidence to support or refute all three of these hypotheses. The problem is in differentiating cause and effect. Insect species diversity, for example, shows a pattern of increasing species richness and abundance from boreal forest through temperate forest types to the humid rain forest, but many other biotic and abiotic environmental parameters are also correlated with this gradient other than primary production. Productivity and insect species diversity may even show an inverse relationship for agro-ecosystems.

4. Predation   Predators and parasites, under non-extreme environmental conditions, are assumed to regulate populations at levels where they avoid intense competition. Paine (1966) carried out a classical study in which the removal of the dominant predator, a starfish species, from a marine littoral community resulted in a decrease in the number of species from 15 to 8. The starfish prevented barnacles and a mussel species from monopolizing space. A similar phenomenon has been described by Estes and Palmisano (1974) where the regulation of sea urchins by sea otters reduced the impact of urchin grazing on the kelp beds and maintained the associated animal community. Harper (1969) suggests that selective grazing by herbivores contributes to the high plant species diversity of grasslands.

5. Spatial heterogeneity   Spatial heterogeneity influences species diversity on a macro- and microscale. Mountainous regions tend to have higher species diversity than comparable lowland areas because a larger number of different habitats are present on an altitudinal gradient. If a comparison of species diversity is made within habitat types, those with a high structural complexity have higher species diversity than more homogeneous habitats. The vertical stratification of a deciduous woodland, for example, provides a complex of habitats for animals which are absent in a mature, evergreen forestry plantation. MacArthur and MacArthur (1961) found a better correlation between bird species diversity and the structural complexity of forests than with plant species diversity in the same habitats. A similar relationship has been described for communities of lizards (Pianka, 1967), rodents (Rosenzweig and Winakur, 1966), meadow invertebrates (Murdoch *et al.*, 1972) and soil animals (Anderson, 1978).

6. Environmental stability Slobodkin and Sanders (1969) developed the environmental stability/time hypothesis which provides a unifying framework into which the other phenomena can be fitted. These authors point out that many environments, such as deserts and estuaries in the tropics have low species diversity despite the latitude and suggest that the combination of the predictability and severity of changes in physical conditions are important determinants of species diversity. In broad terms this relationship does seem to have a descriptive and predictive value. Thus arctic environments show extremes of physical conditions, which are physiologically stressful to animals and plants, and which cannot be avoided by the evolution of appropriate life histories (such as resting stages during periods of drought or extreme cold) because they are unpredictable in time. The organisms in these environments are strongly 'r' selected to rapidly capitalize on favourable conditions so that some individuals are likely to survive severe climatically imposed mortality. Many organisms can adapt to predictable extremes, such as dry seasons or cold winters, and an increase in the predictability of the severe climatic conditions, as in boreal regions, is accompanied by higher animal and plant species diversity. Estuaries are usually occupied by a few abundant species because of the physiological stress associated with the changing freshwater and saline regimes. Estuaries vary in their mixing and circulatory characteristics and greater predictability in the patterns of freshwater flow is reflected by a higher number of marine species being present. Recent exploration of the deep sea trenches has revealed a surprisingly diverse fauna (Wolff, 1977). The great constancy of the physiological environment must contribute to this phenomenon, even though food is limiting to the community. Finally, the physical environment of the humid tropics is both stable, mild and predictable and is associated with high plant and animal species diversity. Equilibrium conditions in the community are, however, never achieved — plants flower at different times, fruits rot and trees fall creating clearings, forming a complex and dynamic mosaic of habitats. The existence of this mosaic over a vast period of time has led to the evolution of these diverse communities.

Trophic organization

Trophic relationships between species may be described in terms of simple food chains or trophic levels, for example the marine food chain: diatom (1° producer) → copepod (1° consumer) → mackerel (2° consumer) → dolphin (3° consumer). Linear food chains rarely occur in natural communities and some parasitoid/insect/host-plant systems are probably the nearest approximation to this situation. Most plants provide food for many different species of herbivore and few predators are prey specific. These trophic relationships are integrated into food webs such as the example shown in Fig. 4.17. This is only part of a larger food web which itself would be much less complex than that for a eutrophic lake community. Even so, it can be seen that there are many problems in classifying some of these animals into their trophic levels. The two caddises, *Dinocras* and *Rhyacophila*, are the top predators consuming the Mayflies as well as the *Hydropsyche* (an omnivore feeding on plant and animal matter). *Perla* is both a top predator and an omnivore. Another stonefly, *Protoneumura*, is entirely saprotrophic, feeding on leaf fragments, while

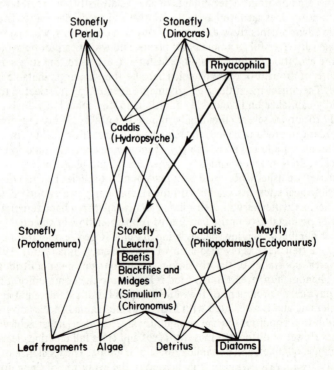

Fig. 4.17 Part of the food web of a freshwater stream in Wales. See text for details. (From Phillipson (1966); after Jones, J. R. E. (1949). *J. Anim. Ecol.,* **18,** 142–59.)

the May-flies and the caddis *Philopotamus* feed on living plants and detritus. An additional complexity is provided by recent research which suggests that aquatic detritivores may be utilizing bacteria on the leaf fragments and not digesting the plant materials themselves.

This complex example is chosen to counteract the impression given in most ecological texts that animal and plant communities can be categorized into three or four trophic levels. The trophic level concept has broad descriptive value for the analysis of ecosystem structure and functioning but as a means of describing community structure it is generally inappropriate, particularly for detritus based communities.

The investigation of feeding relationships in a community is predictably complex. One method which has proved valuable in ecological research is the use of radio isotopes, which are incorporated in animal and plant tissues and distributed through the food web. One of the best examples is a study by Marples (1966) of trophic relationships in a salt marsh community. The vegetation was dominated by the grass *Spartina* which was injected with the isotope ^{32}P. Animals which fed on plant sap, such as aphids, reached a peak loading of isotope first, followed by the general herbivores and then by the predators. Some species, such as the snail *Littorina*, showed very low levels of activity until *Spartina* detritus was labelled. The isotope was then not only located in *Littorina* but also in crabs and even vertebrates such as racoons, which were thus demonstrated to be the top predators of the decomposer subsystem rather than the herbivore subsystem.

The transfer of contaminants through food chains also occurs for several other radio isotopes, such as ^{137}Cesium and ^{90}Strontium, some heavy metals, pesticides and other organic compounds (see Ehrlich *et al.*, 1977 for a comprehensive review of the literature).

In the case of mercury the inorganic compounds are moderately toxic but have a low affinity for living tissues in comparison with the highly toxic organic forms which become concentrated in food chains. Inorganic mercury salts are mobile in the environment and tend to accumulate in marine and lake sediments. It was originally thought that these pools were effectively permanent, but in 1969 it was discovered that bacterial methylation of inorganic mercury was widespread in sediments. Dimethyl mercury is volatile and can cycle through the atmosphere while monomethyl mercury remains in the water body. Methyl mercury is highly neurotoxic to vertebrates and there is concern over the possibility of mercury poisoning from fish similar to the Minamata incident in Japan.

In 1948 an industrial chemical plant was set up in Minamata which used inorganic mercury as a catalyst. Effluents containing mercury passed into Minamata Bay where methyl mercury was concentrated in fish forming a major part of the diet of local villagers. The first patient suffering from neurological damage was diagnosed in 1953. The symptoms of acquired mercury poisoning include sensory disturbances and ataxia (uncoordinated movement) while congenital effects manifest themselves as paralysis, mental retardation and numerous body malfunctions. Although discharges of mercury were stopped in 1960 the cumulative effects of poisoning produced (up to 1972) 292 cases of Minamata Disease including 26 congenital cases and 62 deaths (*Mercury in the Environment*, OECD, Paris 1974). A further outbreak of mercury poisoning from contaminated fish occurred in another region of Japan in 1965 but elsewhere in the world the main sources of fatalities have been the accidental consumption of grain dressed with organic mercury fungicides.

Concentrations of mercury in fish from polluted waters may be up to 10 ppm but most marine fish contain 0.15 ppm, and top predators, such as tuna and sword fish, contain 0.18–2.4 ppm. Mercury poisoning from fish is not considered to be a risk in most areas but fish from some freshwater lakes, such as Lake Erie, and also the sword fish are not considered suitable for human consumption in the U.S.A.

Even where legislation has restricted the use of mercury the recovery of the contaminated waters takes many years.

The chlorinated hydrocarbon pesticides and related compounds, such as the industrial polychlorinated biphenyls, also present a major threat to the biosphere because of their stability (the half life of DDT is more than 10 years) and affinity for lipid tissues. The latter property results in an increase in pesticide concentrations along food chains, which is enhanced, in aquatic systems, by the direct uptake of the pesticide through fish gills. A detailed study by Woodwell *et al.*, (1967) of an estuarine food chain showed a concentration factor from water to fish-eating birds of over half a million times (Table 4.4). As material passes from one link in the food chain to another a major proportion of it is lost by respiration. DDT however is highly soluble in fats which are efficiently transferred between links but represent a small proportion of the respired carbon. The pesticide is thus concentrated in the higher trophic levels. DDT is also absorbed passively by aquatic organisms. A particular property of DDT is that it co-distills with water and has become widespread throughout the biosphere from the polar ice caps to the tropical deep sea trenches. The most obvious manifestations of the effects of organochlorines

Table 4.4 Concentrations of DDT in an estuarine food chain. (Data from Woodwell *et al.*, 1967.)

Organism	DDT concentration (ppm)	Concentration factor with respect to water containing 0.00005 ppm DDT
Phytoplankton	0.4	800
Shrimp	0.16	3200
Predatory fish	1.2–2.1	24 000–42 000
Fish-eating birds	23–26	460 000–520 000

have been on wildlife, particularly birds of prey, and has resulted in legislation against the use of DDT, aldrin and dieldrin in Europe and the U.S.A.
DDT is known to reduce shell thickness in eggs of birds, affect the behaviour of the parents and cause mortality of birds and bats during starvation,
hibernation or migration when the lipid storage reserves are utilized. The
effects of DDT on adult human populations appear to be negligible but
child tolerance and the effects on development have not been monitored
in detail. During the early 1970s in the U.S.A. the milk of many nursing
mothers contained levels of DDT which were unacceptable in commercial
cows' milk. There are numerous documented cases of human poisoning
and deaths attributable to local concentrations of pesticides caused by
mishandling or industrial accidents, but none, to date, through accumulation in food chains.

There are, however, thousands of synthetic compounds in the biosphere
with unknown biological effects. It is only recently that we have become
fully aware of the implications of dioxin poisoning. Dioxin is an impurity
in herbicides containing the compound 2.4.5-T and is one of the most
toxic and mutagenic substances known to man. It now appears that trace
amounts can cause cancer, abortion and genetic deformities in man and
animals. Some of the highest concentrations of dioxin have been found in
the notorious Agent Orange which was used by United States troops to
defoliate forests in South Vietnam. Up to 1971, when spraying stopped,
some 1.5 million hectares (about 10 per cent of the country) was sprayed
at about 13 times the concentration recommended for domestic use. The
ecological consequences of this gross disturbance to natural and agro-
ecosystems are still being assessed (Boffrey, 1971; Shapley, 1973) but
there are indications of severe effects of dioxin on both native populations
and U.S. troops exposed to the spray.

The direct and indirect consequences of altering the trophic structure
of communities is both instructive and cautionary. Zaret and Paine (1973)
have described how the introduction of a predatory fish to Gatun Lake
(Panama) has resulted in changes to a wide range of trophic levels. *Cichla
ocellaris*, a voracious predator from the Amazon River, is a prized sporting
fish and excellent eating. It was accidentally released into Gatun Lake,
though other Central American lakes were deliberately stocked with the
species. *Cichla* bred rapidly and had a devastating effect on seven out of
the eight common fish species; six species were effectively eliminated
resulting in a gross simplification of the food web. The indirect consequences of these changes were a reduction in tern, kingfisher and heron
populations in the region, changes in zooplankton density (consequent
changes in phytoplankton production are predicted) and indications that
a resurgence of local mosquito populations, which are malarial vectors,
may have been caused by the reduction in insectivorous fishes.

This example contrasts the effects of predators in equilibrium communities discussed above.

Succession

Succession cannot be directly observed unless communities are disturbed in some way. If a forest is cleared for farming, it will normally revert to forest following the cessation of agricultural practice. An area of bare rock, or a disused road, will be colonized by lichens and mosses, then by grasses and herbs and eventually, under favourable conditions, by woody perennials. A shift in the river channel of an estuary may erode one region and deposit silt in another. The silt will be first stabilized by salt-tolerant marsh vegetation and then grasses and shrubs will become established as the soil level rises and salt is leached out. All these communities show a pattern of development towards a more mature stage of development, the *climax*, which is characteristic (and therefore predictable) for particular environmental conditions.

The characteristics of a community are determined by the relationship between the organisms and their physical and chemical environment. If any of these parameters act positively within the environmental complex then the community will develop, while a negative balance will cause recession or degradation. Silting induces succession in salt marsh vegetation while erosion will reverse this process. Eutrophication of lakes by an exogenous nutrient source induces maturation over time while nutrient leaching from a soil can limit the development of a plant community, for example in heathlands. Change induced by these exogenous variables is known as *allogenic succession*. *Autogenic succession* is induced by positive feedback within the community (endogenous variables) such as nitrogen fixation, the infilling of lakes by organic matter (bog lakes) rather than by silt, or the drying out of soil by transpiration.

Bog lakes are characterized by floating mats of vegetation, often dominated by the moss *Sphagnum*, which grow out towards the centre of the lake, cutting off the open water and infilling the lake basin with peaty organic matter. The moss mat is consolidated by grasses, herbs, shrubs and moisture tolerant trees, such as willows and alder, and is eventually converted into dry land. Many lakes show infilling by both allogenic and autogenic processes and a succession sequence, or *hydrosere*, can often be observed from open water to dry land representing the different phases of successional development.

Autogenic succession is rarely observed in detail because of the time span involved. A well documented example, however, is the post-glacial succession at Glacier Bay, Alaska (Crocker and Major, 1955, Lawrence *et al.*, 1967), which illustrates the interaction between plants inducing and regulating successional change. The recession of glaciers has occurred over about 100 kilometres since 1750 and has left moraines* devoid of vegetation. The exposed till (boulder-clay) is colonized by mosses and two or three shallow rooting herbacious species. Willows, prostrate species at

*A moraine is a continuous marginal line of debris borne on or left by a glacier.

first, then shrubby species, become established over about 15 years or more. Alder quickly invades once the shrubby species are established and forms thickets up to 10 m tall after about 50 years. The alder is invaded by sitka spruce, forming a dense forest in about 150 years, which continues to mature as hemlock invades to form a climax after 200 years or more. In areas of poor drainage the forest may be invaded by *Sphagnum* which holds the water, causes waterlogging of the soil and kills the trees. The climax in these areas is a bog.

One of the principle phenomena governing the rate of succession and development of the above community is the accumulation of a large nitrogen capital. *Dryas*, a herbacious species characteristic of the early successional phases, and alder have nitrogen-fixing symbionts which result in an increase in soil nitrogen over the pioneer phase (Fig. 4.18). Alder also acidifies the soil so that the spruce is able to invade and displace the alder using the accumulated nitrogen capital. The soil nitrogen content falls in the mature stages of succession as nitrogen is converted into tree biomass.

The climax vegetation of a region is primarily determined by climate (see Fig. 3.1) but topography, soil type, fire or grazing can result in the formation of a mosaic of different plant communities within a region. A long-term change in the prevailing climatic conditions, or a short-term change in local conditions, will result in a shift to a new equilibrium condition. Following the reduction of rabbit populations by myxomatosis

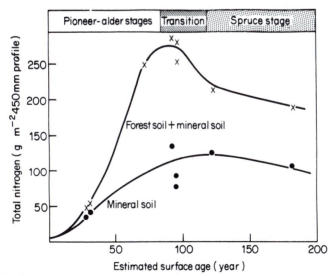

Fig. 4.18 Total nitrogen content of soils recently uncovered by glacial retreat at Glacier Bay, Alaska. Plant succession is shown at the top of the diagram. (After Crocker and Major, 1955.)

in the United Kingdom, many grassland areas showed a change in plant species composition, particularly an increase in tree seedlings which had previously been grazed by the rabbits (Thomas, 1963). Today conservational measures are required to prevent the succession in these areas from moving towards a woodland climax, which would exclude many rare plant species. Agro-ecosystems are similarly prevented from maturing by the energy input of agricultural practice.

Some degree of change in plant species distribution and abundance is characteristic of all climax communities, whether induced by land slips, tree falls or by cyclical characteristics of the vegetation. One of the best documented cyclical changes is in the heather *Calluna vulgaris* which dominates many heathlands in Northern Europe. If a *Calluna* stand is undisturbed by burning or grazing, or is not overgrown by trees, the plants eventually begin to die naturally of old age. As the plants degenerate the branches spread out and die back forming a progressively widening gap in the centre of the patch which is colonized by *Calluna* seedlings and other plants. Eventually an uneven aged stand is formed of *Calluna* plants in various stages of the cycle. In this example the cycle is induced by the growth characteristics of the vegetation but cyclical changes, on much larger scales in space and time, can also be induced in forest stands by climatically controlled effects. Miles (1979) provides an excellent review of these and other aspects of succession, cyclical changes and other dynamic processes in vegetation communities.

Odum (1969) has summarized the characteristics of developmental and mature stages of ecosystems (Table 4.5) which integrates many of the points made in previous chapters. While most of these ecosystem properties are generally accepted, at least for successions leading to forest, a number are arguably correct for many situations. For example, in number 7 in the table, nitrogen is accumulated during succession but from the atmosphere and not from the soil; further, precipitation is the primary source of phosphorus in some ecosystems. The ratio of minerals in soil and vegetation also varies considerably with ecosystem type and latitude. A second point relates to species diversity (numbers 8 and 9). Climax ecosystems are not necessarily diverse compared with the early successional stages. Open shrubby vegetation may contain far more animal and plant species with greater equitability than a pine forest climax on the same site. Finally, the resistance of ecosystems to perturbations (number 22) is not necessarily concomitant with their maturity. Tropical rain forests are among the oldest and most species diverse ecosystems but are extremely fragile and easily perturbed by agricultural and forestry practices. Even local cutting of the trees destroys the humid microclimate of the forest, the soil is exposed to extreme climatic conditions and the nutrient capital, accumulated over thousands of years, may be lost as the soil is eroded. Many deforested regions of the humid tropics are unlikely to return to the original vegetation community even if they remained undisturbed for any conceivable scale of time.

Table 4.5 A tabular model of ecological succession: trends to be expected in the development of ecosystems. (From Odum, 1969.)

Ecosystem attributes	Developmental stages	Mature stages
Community energetics		
1. Gross production/community respiration (P/R ratio)	Greater or less than 1	Approaches 1
2. Gross production/standing crop biomass (P/B ratio)	High	Low
3. Biomass supported/unit energy flow (B/E ratio)	Low	High
4. Net community production (yield)	High	Low
5. Food chains	Linear, predominantly grazing	Weblike, predominantly detritus
Community structure		
6. Total organic matter	Small	Large
7. Inorganic nutrients	Extrabiotic	Intrabiotic
8. Species diversity – variety component	Low	High
9. Species diversity – equitability component	Low	High
10. Biochemical diversity	Low	High
11. Stratification and spatial heterogeneity (pattern diversity)	Poorly organized	Well-organized
Life history		
12. Niche specialization	Broad	Narrow
13. Size of organism	Small	Large
14. Life cycles	Short, simple	Long, complex
Nutrient cycling		
15. Mineral cycles	Open	Closed
16. Nutrient exchange rate, between organisms and environment	Rapid	Slow
17. Role of detritus in nutrient regeneration	Unimportant	Important
Selection pressure		
18. Growth form	For rapid growth (r-selection)	For feedback control (K-selection)
19. Production	Quantity	Quality
Overall homeostasis		
20. Internal symbiosis	Undeveloped	Developed
21. Nutrient conservation	Poor	Good
22. Stability (resistance to external perturbations)	Poor	Good
23. Entropy	High	Low
24. Information	Low	High

5 Primary Production

Discussion of primary production in the ecological literature is generally concerned with net primary production (NPP). This is because NPP is not only a measure of energy or matter potentially available to heterotrophs, including man, but also because direct measurements of NPP can be made more readily than for gross primary production (GPP) or photosynthesis; particularly in terrestrial ecosystems. Methods for determining NPP are summarized by Newbould (1967) for forests, Milner and Hughes (1968) for grasslands and Vollenweider (1969) for aquatic systems.

Primary production, even under optimal growth conditions, represents only a small fraction of the photosynthetic active radiation (PAR) potentially available to plants. The efficiency of photosynthesis and the amount of autotroph respiration both substantially reduce the energy flux to higher trophic levels.

Photosynthesis and Respiration

The amount of solar radiation reaching the earth's atmosphere (the solar constant) is 1358 W m^{-2}. As the radiation passes through the atmosphere energy is lost through absorption and scattering (p. 8) so that only 910 W m^{-2} (a world average) reach ground level. Of this 420 W m^{-2} are in the PAR wavelengths (400–700 nm) and about 85 per cent of this may be absorbed by plants. As much as 95 per cent of the absorbed energy is lost as heat while less than 5 per cent is captured by photosynthesis.

The efficiency of photosynthesis can be calculated by expressing GPP as a percentage of PAR. Forests have the highest gross production (photosynthetic) efficiencies of 2.0–3.5 per cent while herbaceous communities, including agricultural crops, are in the order of 1–2 per cent. Values for phytoplankton communities are usually less than 0.5 per cent.

The reasons for the differences in GPP or photosynthetic efficiency between different types of plant and different plant communities are complex and involve the interaction of many physiological and ecological factors.

The major physiological determinants of photosynthetic rates for individual plant leaves are light, CO_2 availability and moisture. Underlying the response of plants to these variables are two main types of photosynthetic process: the C_3 and C_4 photosynthetic pathways.

In *C_3 plants* the primary products of CO_2 fixation are 3-carbon acids derived from the carboxylation and cleaving of a 5-carbon acceptor molecule:

$$\text{ribulose diphosphate (5C)} + CO_2 \rightarrow 2 \text{ phosphoglyceric acid (3C)}$$

This carbon fixation reaction (part of the Calvin cycle) is the most widely distributed type of photosynthesis and occurs in all algae and most vascular plants.

In *C_4 plants* the carboxylation of a 3-carbon acceptor molecule results in the formation of 4-carbon organic acids (Hatch and Slack pathway):

$$\text{phosphoenolpyruvic acid (3C)} + CO_2 \rightarrow \begin{array}{l} \text{aspartic acid} \\ \text{malic acid} \end{array} \text{(4C)}$$

The 4-carbon acids are decarboxylated and the CO_2 released is fixed by the Calvin cycle.

Both monocots and dicots have C_3 and C_4 representatives but the C_4 pathway is less common, occurring only in vascular plants, and about half of the C_4 species are grasses. Some of the characteristics of C_3 and C_4 plants are summarized in Table 5.1

Table 5.1 Some photosynthetic characteristics of C_3 and C_4 plant species. (Based on Black, C. C. (1973). *Annual Review of Plant Physiology*, **24**, 253–86.)

Characteristic	C_3	C_4
Leaf anatomy	No distinct bundle sheath of photo-synthetic cells	Well organized bundle sheath
Carboxylating enzyme	Ribulose diphosphate carboxylase (RuDP)	PEP carboxylase then RuDP
CO_2 compensation point (ppm CO_2)	30–70	0–10
Photosynthesis inhibited by 21 per cent O_2?	Yes	No
Photorespiration detectable?	Yes	Only in bundle sheath
Optimum temperature for photosynthesis (°C)	15–25	30–40
Transpiration ratio (g H_2O g dry wt increase^{-1})	450–950	250–350
Dry matter production (t ha^{-1} yr^{-1})	22 ± 0.3	39 ± 17

Light effects

The light intensity at which photosynthesis balances respiration is called the *light compensation point* and growth can only occur above this level. The compensation point varies with C_3 or C_4 species, light intensity, temperature and CO_2 concentration but is usually less than 2 per cent of full sunlight. Differences in light compensation points are caused primarily by differences in respiration rates: when respiration is low the leaf requires less light for CO_2 fixation than when respiration is high.

Figure 5.1 shows the light intensity response of single leaves from plants grown in contrasting habitats. The upper curve is for *Tidestromia*, a C_4 shrub which grows under extremely hot and dry conditions in a Californian desert (Death Valley), the middle curve is for *Atriplex* growing in a mesic environment along the Pacific coast of the U.S.A. and the lower curve is for a rain forest herb, *Alocasia*, from Northern Australia. The PAR energy received each day by *Tidestromia* and *Alocasia* varies 300 fold (Salisbury and Ross, 1978). The response of *Alocasia* is typical of shade plants which exhibit much lower photosynthetic rates, even under bright sunlight, than species from open areas. Conversely, shade plants are able to photosynthesize at higher rates under low light intensities because their light compensation points are low. These characteristics enable them to grow slowly in their natural habitats where other species, with higher compensation points would die.

The light response of *Tidestromia* is typical of C_4 species from sunny habitats, including crops such as corn (maize), sorghum and sugar cane

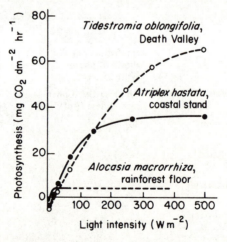

Fig. 5.1 Influence of light on photosynthetic rates in single attached leaves of C_3 and C_4 dicot species from different habitats. (From Salisbury, F. B. and Ross, C. W. (1978). *Plant Physiology*, second edition. © 1978 by Wadsworth Publishing Company, Inc., Belmont, California 94002. Reprinted by permission of the publisher.)

which show no rate saturation up to and even beyond the intensity of full tropical sunlight. Peanut and sunflower, however, are two C_3 plants which show a light saturation response almost as high as C_4 species. Most tree species are intermediate between the typical C_3 species and shade plants and are often saturated at less than 25 per cent full sunlight.

Photorespiration

It has been found that respiration in C_3 species may be 2 or 3 times higher in bright sunlight than in darkness and can be half the net CO_2 fixation rate. This effect is not caused by an elevation of normal cellular respiration in cytoplasm and mitochondria but by *photorespiration*, causing a loss of fixed CO_2 from the chloroplasts. Photorespiration is virtually undetectable in C_4 species and accounts for their much higher photosynthetic rates at high light intensities (Table 5.1). Photorespiration is also stimulated by high temperatures, low CO_2 levels and high O_2 levels. In fact it was the inhibition of algal photosynthesis by oxygen, noted by Warburg, which directed attention to the phenomenon. The elimination of photorespiration from C_3 plants by genetic engineering would significantly increase their productivity.

Carbon dioxide and moisture effects

During the main growing season lack of available CO_2 within the leaves of plants is a common cause of sub-optimal photosynthetic rates. The C_3 and C_4 plants, however, respond differently to low CO_2 tensions. At high light intensities the balance between CO_2 fixation and respiration (the CO_2 compensation point) usually reaches zero in C_3 plants at 50–100 ppm CO_2 but is still positive at 0–5 ppm CO_2 in C_4 plants. The low CO_2 compensation point of C_4 plants arises from the much lower photorespiratory release of CO_2 caused by high light intensities. The consequence of this is that under conditions of moisture stress when the stomata in the leaves begin to close, or under arid conditions where the leaves are adapted to reduce moisture losses, CO_2 uptake is restricted and photosynthesis in C_3 plants becomes limited. The C_4 plants, however, can continue at high rates of photosynthetic activity and control transpiratory moisture losses resulting in a more efficient use of available water (Table 5.1).

Net primary production

The percentage of gross primary production expended by autotroph respiration is related primarily to the plant biomass which the photosynthetic tissues have to support. Thus in phytoplankton communities respiration is 30–40 per cent of GPP but increases to about 40 per cent for temperate grassland and agricultural crops, 50–60 per cent for temperate forests and to 70–80 per cent for tropical forests (Odum, 1969).

The overall efficiency of energy fixation by plants is expressed by the net production efficiency (NPP divided by PAR). Various estimates of the theoretical upper limit to plant yield have been made, ranging from 2–5 per cent (Bonner 1962) to 10 per cent (Thornley 1970), but observed efficiencies are generally far below this level. Phillipson (1973) calculated that the mean production efficiencies for 46 woodland and 57 non-woodland ecosystems were 0.54 per cent and 0.58 per cent respectively. Knowing the input of PAR to the land surface, the theoretical and observed levels of NPP can be compared. Figure 5.2 shows the comparison of natural and agro-ecosystems with the theoretical units of production using a net production efficiency of 2.6 per cent (assuming

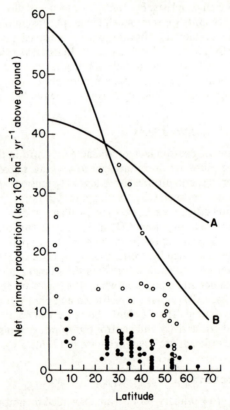

Fig. 5.2 Net primary production in non-woodland ecosystems. Curve **A** is based on a net photosynthetic efficiency of 2.6% and the assumption that 22% of the extra-terrestrial solar radiation reaches the ground. Curve **B** is also based on a net photosynthetic efficiency of 2.6% but allows for the mean annual prevelance of cloudiness at different latitudes. Closed circles depict natural and semi-natural systems, open circles depict man-made ecosystems. (Redrawn from Phillipson, 1973.)

in this case that 22 per cent of extra-terrestrial radiation reaches the ground) and allowing the mean annual prevalence for cloudiness for each latitude. In a few cases the agricultural systems reach the theoretical maximum but in most cases they do not and natural ecosystems are only 10–20 per cent of this value. The same pattern is evident for the forest systems. Clearly light is only one of the factors limiting terrestrial productivity and maximum NPP is only attained under optimum conditions of light, temperature, moisture and mineral nutrients. Optimization of all these conditions are rarely achieved, even by intensive agriculture.

Global Patterns of NPP

Estimates of world NPP are somewhat variable between authors but most values fall within the range $100–300 \times 10^9$ tonnes dry matter (equivalent to 3.35×10^{21} J); a recent value of 176.7×10^9 tonnes has been calculated by Leith (1975), of which 33 per cent is contributed by the oceans and 67 per cent by the continents. It is evident from this ratio, which is almost the exact reverse of the relative areas of sea and land on the surface of the globe, that rates of primary production are not evenly distributed in the biosphere. This is emphasized by the breakdown of total NPP into the contribution by the major terrestrial and aquatic ecosystem types shown in Table 5.2.

There is a broad latitudinal trend in NPP from the tundra (140 g m^{-2} yr^{-1}) through the temperate zones ($600–1200$ g m^{-2} yr^{-1}) to the tropics ($900–2200$ g m^{-2} yr^{-1}). This follows general trends in rainfall, mean annual temperature and plant biomass so that any of these parameters can be used to predict NPP in a particular region. Water is a major determinant of NPP in terrestrial ecosystems and production shows an almost linear increase to increases in annual precipitation for arid regions. In high rainfall areas, such as the humid tropics, production shows little further increase over about 2200 mm per annum. The production supported by a given level of precipitation depends, however, on the temperature conditions under which rainfall occurs i.e. the evapotranspiration characteristics of a region. The correlation between evapotranspiration (E) and NPP (P) is consequently much higher than for rainfall alone and is described by the function:

$$P = 3000 \left(1 - e^{-0.0009695 \, (E - 20)}\right)$$

Leith and Box (1972) used this model to convert a computer-simulated world map of actual evapotranspiration into an NPP map which showed close approximation to observed patterns of production.

Oceanic NPP is essentially limited by a combination of light intensity and nutrient availability according to latitude and local conditions. There are few areas which approach optimization of both of these parameters.

Table 5.2 Net annual primary production and plant biomass for the Earth (dry weight). (From Whittaker, 1975.)

Ecosystem type	Area (10^6 km^2)	Net primary productivity, per unit area (g m^-2 or t km^-3) Normal range	Mean	World net primary production (10^9 t)	Biomass per unit area (kg m^-2) Normal range	Mean	World biomass (10^9 t)
Tropical rain forest	17.0	1000–3500	2200	37.4	6–80	45	765
Tropical seasonal forest	7.5	1000–2500	1600	12.0	6–60	35	260
Temperate evergreen forest	5.0	600–2500	1300	6.5	6–200	35	175
Temperate deciduous forest	7.0	600–2500	1200	8.4	6–60	30	210
Boreal forest	12.0	400–2000	800	9.6	6–40	20	240
Woodland and shrubland	8.5	250–1200	700	6.0	2–20	6	50
Savanna	15.0	200–2000	900	13.5	0.2–15	4	60
Temperate grassland	9.0	200–1500	600	5.4	0.2–5	1.6	14
Tundra and alpine	8.0	10–400	140	1.1	0.1–3	0.6	5
Desert and semidesert shrub	18.0	10–250	90	1.6	0.1–4	0.7	13
Extreme desert, rock, sand, and ice	24.0	0–10	3	0.07	0–0.2	0.02	0.5
Cultivated land	14.0	100–3500	650	9.1	0.4–12	1	14
Swamp and marsh	2.0	800–3500	2000	4.0	3–50	15	30
Lake and stream	2.0	100–1500	250	0.5	0–0.1	0.02	0.05
Total continental	149		773	115		12.3	1837
Open ocean	332.0	2–400	125	41.5	0–0.005	0.003	1.0
Upwelling zones	0.4	400–1000	500	0.2	0.005–0.1	0.02	0.008
Continental shelf	26.6	200–600	360	9.6	0.001–0.04	0.01	0.27
Algal beds and reefs	0.6	500–4000	2500	1.6	0.04–4	2	1.2
Estuaries	1.4	200–3500	1500	2.1	0.01–6	1	1.4
Total marine	361		152	55.0		0.01	3.9
Full total	510		333	170		3.6	1841

Mean NPP of oceanic regions, about 90 per cent of the total marine area, is in the order of $125 \ g \ m^{-2} \ yr^{-1}$; a figure comparable with the tundra or semi-desert conditions on land.

The oceans are permanently stratified from a latitude of about $45°N$ to about $45°S$ and within this region nutrient levels in the euphotic zone are extremely low. Phytoplankton sinks at mean rates ranging from about 3 m day^{-1} in cool waters to 6 m day^{-1} in warmer waters. Nutrients in cells sinking below the thermocline are effectively lost from that region of the water column but grazing, excretion, and bacterial decomposition are important nutrient recycling processes in surface waters. The rate of turn-over of nutrients within the plankton system largely determines steady state levels of oceanic NPP; limited nutrient return occurs through horizontal currents in the open oceans.

The stratification breaks down in polar regions where cold water $(0-5°C)$ extends to the surface. Here the primary limitation to NPP is low light intensity, particularly during the winter months, with nutrient availability imposing secondary effects. In some areas of the tropics upwelling of nutrient rich waters occurs where deep currents flow along continental masses and the surface waters are displaced by offshore winds. Under these conditions NPP is high $(400-1000 \ g \ m^{-2} \ yr^{-1})$ in comparison with other oceanic regions and, although the areas of upwelling are small, they are associated with some of the most productive marine fisheries.

The inshore regions of the continental shelves are shallow (the dome of St. Paul's Cathedral would not be submerged in most areas of the North Sea) and well mixed by wave action and tidal currents. Domestic sewage and agricultural runoff contribute significantly to the eutrophication of inshore waters (Ryther and Dunstan, 1971) and NPP ranging from $200-600 \ g \ m^{-2} \ yr^{-1}$, may be compared favourably with the upwelling zones. The net production efficiency is only about 0.07 per cent, as a result of the high extinction coefficient of turbid inshore waters. Productivity can also be limited, however, by a high mixing rate in unstratified waters with shallow euphotic zones. Under these conditions the density of photo-synthesizing plankton may be diluted by waters from below the compensation level.

Phosphorus and nitrogen commonly limit primary production in the seas but other nutrients may become limiting locally or under particular conditions. In the Sargasso Sea, for example, iron is marginally more limiting to production than nitrogen or phosphorus (Menzel and Ryther, 1961).

The potential productivity of aquatic and terrestrial systems may be compared in terms of the standing crop of available nutrients. Rich fertile soil contains up to 0.5 per cent available nitrogen which can support $50 \ kg \ m^{-2}$ (dry weight) of vegetation. The richest oceanic waters contain 0.5×10^{-4} per cent nitrogen and no more than $5 \ g \ m^{-2}$ (dry weight) of phytoplankton (Ryther, 1969). It is the vast area of the oceans and the high turnover rates of the populations in the oceans which provide the large proportional contribution of marine NPP to total production. Man does not, however, significantly exploit marine primary production, as on land, but capitalizes on energy and nutrients converted to the secondary production of the higher trophic levels.

The Effects of Pollutants on Primary Production

Many minerals and industrial compounds present in soils at high concentrations affect the growth of terrestrial vegetation. The effects are generally localized around the source and are primarily of significance in the reclamation of industrial land such as mine and colliery waste tips (see Chadwick and Goodman, 1975). Atmospheric pollutants can have chronic effects far beyond the point of emission. Concern has primarily been directed towards the effects of ozone (O_3) and oxides of sulphur and nitrogen.

Ozone causes severe damage to plants (chlorosis and necrosis) at concentrations as low as 15 pphm (parts per hundred million) when exposure continues over long periods of time. Recently synergistic effects (interactions where the total effect is greater than the sum of the individual effects) between O_3 and SO_3 have been identified which occur at levels far below the thresholds of the individual gasses; as low as 5 pphm O_3 for some plant species (Williams and Ricks, 1975). Mixtures of O_3, SO_2, NO_2 and the highly toxic phyto-oxidant PAN are found in photochemical smogs. Smogs originating from Los Angeles have caused moderate to severe damage to trees over some two-thirds of the 65 000 km^2 La San Bernadino National Forest, 125 km down-wind of the city (E. Hay in Ehrlich *et al.*, 1977). Similarly, SO_2 emissions from an iron foundry in Canada caused total destruction of vegetation 8 km down-wind, considerable damage to trees and shrubs up to 16 km away and detectable damage at more than 30 km (Gordon and Gorham, 1963).

Other less obvious, but more widespread effects of atmospheric pollution by SO_2 and NO_2 are manifested in acid rains. Rain pH values below 4, and exceptionally around pH 2.0, have been recorded in the north-eastern U.S.A. and northern Europe. In Scandinavia the effects of acid rain, partly caused by industrial emissions from the U.K. and Germany, include increased leaching of cations from soils and leaf surfaces, changes in the structure of terrestrial and aquatic plant communities and the destruction of salmon and trout fisheries (Braekke, 1976). This report found no evidence of earlier claims that coniferous forest productivity and nitrogen fixation was affected, but the effects on base rich soils are largely unknown.

Plant species vary considerably in their tolerance of SO_2 pollution. The lichens and mosses are generally more susceptible than higher plants and the species occurrence and growth forms of these groups have been used as atmospheric pollution indicators. Creed *et al.* (1973) have shown that the mean concentration of SO_2 (mg m^{-2}) can be determined from the reflectance of trees at 1.5 m, the mean height of the highest moss above the ground and the percentage cover of foliose lichens:

$$\sqrt{SO_2} = 18.72 - 3.94 \text{ reflectance} - 0.15 \text{ height} - 2.38 \text{ foliose}$$

The regression accounted for 71 per cent of the variance and hence has a high predictive value for the English counties. The constants would have to be established empirically for other regions.

Heavy metals, particularly copper (Cu) and mercury (Hg), and chlorinated hydrocarbons have been found to reduce photosynthetic activity in marine and freshwater phytoplankton; as well as some macrophytic algae. In one study production by marine phytoplankton was reduced to 50–80 per cent of controls by 10 ppb DDT but no effects were detected below 1 ppb (Wurster, 1968). Mosser *et al.* (1972) demonstrated similar effects of PCB (polychlorinated biphenyls) concentrations as low as 10–25 ppb on two species of algae while two other species were unaffected. Mercury also reduced the photosynthetic activity of a marine diatom and freshwater plankton samples at levels (as organomercury fungicides) of only 0.1–0.5 ppb (Harriss *et al.*, 1970). A large number of pesticides and heavy metals have been shown to have broad spectrum or differential effects on phytoplankton but generally at levels below those encountered in natural environments. DDT, for example, was detected in concentrations as low as $2.3-5.6 \times 10^{-12}$ g m^{-1} in sea water off the Californian coast but as high as 5.7×10^{-6} g g carbon^{-2} (5.7 ppm) in fine suspended particulate matter (Cox, 1971). Cox suggested that over 90 per cent of DDT in water may become bound to these particles (the actual solubility of DDT in water is only 1.2 ppb) where it is unavailable for uptake by algae. Recently, however, Harding and Phillips (1978) have found that while PCBs are hydrophobic and, like DDT (and mercury), become particle bound, the adsorbed materials can be taken up by phytoplankton and move to photosynthetically active sites. Recurrent concern over the possible effect of pollutants on marine primary production is also compounded by the general absence of information on synergisms between pollutants.

At the present time there is no conclusive evidence of pollutants inhibiting either terrestrial or aquatic production on a global scale and, as discussed earlier, eutrophication may enhance primary production. There is more evidence accumulating however that both biodegradable and non-degradable pollutants can cause changes in plant community structure below levels at which total NPP is affected. The consequences of these changes for ecosystem functioning are unknown.

World Crop Production

Although some 80 species of food plants have been selectively cultivated over the centuries, 56 per cent of present world food production is represented by cereals, of which over 40 per cent is rice and wheat (Table 5.3).

Table 5.3   Sources of food energy for man (From Ehrlich *et al.*, 1977.)

Food	Percentage of energy supplied
CEREALS	56
Rice	21
Wheat	20
Corn	5
Other cereals	10
ROOTS AND TUBERS	7
Potatoes and yams	5
Cassava	2
FRUITS, NUTS AND VEGETABLES	10
SUGAR	7
FATS AND OILS	9
LIVESTOCK PRODUCTS AND FISH	11
TOTAL	100

Cereals form approximately 50 per cent of both the energy and protein consumed by the human population of the world. Rice is the staple diet for over 2000 million people and total world production in 1974 was estimated as 323 million tonnes. Rice is principally cultivated in the tropics with China and India, the largest single producers, contributing 5 per cent and 22 per cent of the world total respectively.

Wheat, however, is principally cultivated in temperate regions where cold, wet winters and hot, dry summers discourage fungal pathogens; particularly wheat-rust which thrives under humid conditions. World production of wheat was some 360 million tonnes (in 1974) of which the Soviet Union produced almost 30 per cent, and the U.S.A. and China 10–12 per cent each.

Maize production in the same years was in the order of 293 million tonnes. This figure is not proportionately represented in Table 5.3 because the bulk of U.S. production, 45 per cent of the world total, is used as animal feed.

The remaining cereals, barley, oats, rye, millet and sorghum, contribute slightly more than 50 per cent of the world total cereal production of approximately 2 billion tonnes. More than half of this total production comes from the U.S.A., U.S.S.R. and western Europe.

Potato production is approximately 300 million tonnes. As this represents only 75 million tonnes dry matter, the contribution to world food production is small but, nevertheless, important because potatoes have high protein content.

Legume production was 124 million tonnes in 1973 and provided about 20 per cent of the world's protein supply. Soy beans and groundnuts, while high in protein content, are principally used as vegetable oils and animal feeds.

These figures mean little in an abstract context because world food resources are not partitioned on a *per capita* basis. Average human requirements for protein and energy are shown in Fig. 5.3 in relation to the diets of people in the highly developed and less developed regions of the world.

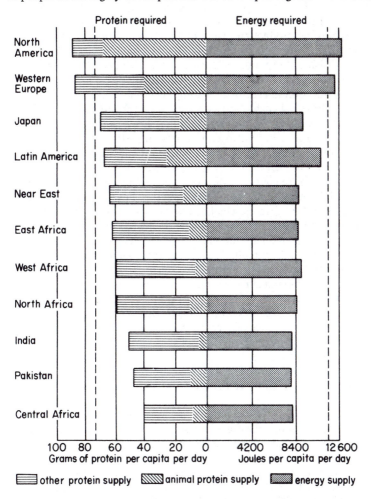

Fig. 5.3 Daily protein and energy requirements and average intake for populations in major regions of the world. (From *Provisional Indicative World Plan for Agricultural Development* (1970). U.N. Food and Agriculture Organization, Rome.)

Most people in North America and Western Europe exceed optimum nutritional levels and medical concern over obesity in these countries contrasts starkly with most other areas of the world where insufficient food is available. No one knows quite how many people are malnourished (lacking in nutrients, particularly protein) or undernourished (insufficient energy in the diet) but it is generally accepted that two-thirds of the world population is receiving a sub-optimal diet, with perhaps 500 million people at the starvation level.

There was widespread optimism in the late 1960s that improvements in world agricultural technology, particularly in the less developed regions, would allow food production to exceed the 2 per cent annual growth of the human population. The optimistic forecasts were based on the development of high yield varieties of cereals (HYVs) which can produce up to four times the yield of traditional varieties. A particular property of the HYVs was the development of a dwarfing gene which gives the plant a short, stiff straw and enables it to respond more proportionately in yield to fertilizer application. Traditional varieties of rice and wheat cannot make efficient use of high nitrogen fertilizer, in particular, as it causes them to grow too tall and topple over. Other genetic improvements include resistance to microbial pathogens, but pesticides need to be applied in large quantities to control insects in monocultures of such high quality food. In addition to fertilizer and pesticides the HYVs respond best to controlled supplies of water, which necessitates irrigation rather than a reliance on rainfall – about 14 per cent of the world's agricultural land is irrigated. Finally, rapid maturation and insensitivity to day-length in the new seed varieties allows multiple cropping in favourable regions. In some wet areas of India, three main crops can be harvested every 14 months while in the more seasonal regions a summer wheat crop can be alternated with a winter crop of rice.

The initial effect of HYVs on world agricultural production was dramatic. India's wheat harvest more than doubled in 6 years and there were large gains in cereal production in Pakistan, the Phillipines and Sri Lanka. Subsequently yields have declined but a baseline for production by current HYVs, grown under the same conditions as traditional varieties, is about 50–100 per cent increase for wheat and 10–25 per cent for rice (Dalyrymple, 1975).

In 1972, after nearly two decades of increases in world food production, adverse weather conditions affected most of the major grain producing regions of the world and there was a short fall of 3 per cent in grain harvest; 33 million tonnes below the 1971 harvests at a time when 25 million tonnes more than the previous year were required. World grain reserves were sufficient to support the short fall but prices soared. Between 1971 and 1973 the average prices of wheat on the world market rose by 124 per cent, rice by 186 per cent, maize by 69 per cent and soy beans by 130 per cent. Massive purchases of wheat from the U.S.A. by

Russia did much to increase prices on the world market. Higher food prices inevitably affected the poorest people. The grain stocks were not built up in 1973 although there was a good harvest, and then another bad crop followed in 1975. On this occasion bad weather coincided with the energy crisis and consequent fertilizer shortages. The short fall in production amounted to some 50 million tonnes; a drop of more than 4 per cent below 1971 levels. World grain reserves fell at one point to a 33 day supply. The food crisis of 1972–1974 is believed to have killed at least 2 million people in Bangladesh, Ethiopia, the African Sahel and three states of India alone.

There have been good and bad harvests since then but the significance of this unpredictable sequence of poor harvests is that world food production, notwithstanding technological developments, is still highly susceptible to adverse weather conditions. The productive years of the late 1960s, on which many optimistic forecasts of food production were based, were an unusually favourable period of global weather conditions.

Land is the most essential requisite for crop production. Recent studies suggest that there are at most 3.2 billion hectares of land potentially suitable for agriculture. Approximately half of that land, the richest and most accessible land, is already under cultivation and the remainder requires intense capital and energy inputs to bring it into production. The world average of agricultural land *per capita* is only 0.4 ha and this will be reduced to 0.25 ha or less if the world population increases to the higher levels of 10–16 billion predicted by the year 2100 (Pimentel and Pimentel, 1979). This estimate assumes that the area of agricultural land remains constant, but in the last 200 years at least a third of top soil on U.S.A. croplands has been lost through erosion. Some 4 billion tonnes of sediments are washed into U.S.A. waterways each year, of which 75 per cent comes from agricultural land (Pimentel *et al.*, 1976). Land degradation is more rapid elsewhere in the world where marginal land, often on steep slopes, is being pressed into production and desertification continues, often as a result of overgrazing. As human populations expand, further reductions in agricultural land are imposed by housing, roads and industrial development (Fig. 5.4).

To feed the world population, projected to increase by 6 or 7 million in 25 years, world food production must be doubled on currently available land. This will require a 3-fold increase in energy input to agricultural systems (approximately 50 per cent as fertilizers and 50 per cent as fuel). The developing countries already allocate more than 60 per cent of their energy to food production compared with only about 16 per cent in the United States and United Kingdom; on a world wide basis this proportion is nearly 25 per cent (Pimentel and Pimentel, 1979).

Much of modern agriculture is highly energy subsidized, i.e. the energy content of products is much lower than the input of fuels, fertilizers and feeds as well as the costs of their production and distribution. Comparisons

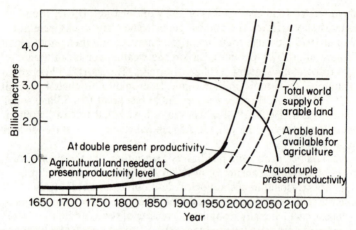

Fig. 5.4 A theoretical prediction of future relationships between human populations, food production and the availability of agricultural land. Total world supply of arable land is about 3.2 billion hectares. About 0.4 hectares per person of arable land are needed at present productivity. The curve of land needed thus reflects the population growth curve. The thin line after 1970 shows the projected need for land, assuming that world population continues to grow at its present rate. Arable land available decreases because arable land is removed for urban-industrial use as population grows. The dotted curves show land needed if present productivity is doubled or quadrupled. (From Meadows, D. H., Meadows, D. L., Randers, J. and Behrens, W. W. *The Limits to Growth: A Report for The Club of Rome's Project on the Predicament of Mankind*. A Potomac Associates book published by Universe Books, New York, 1972, 1974. Graphics by Potomac Associates.)

of different farming systems around the world in terms of energy input/ output ratios are shown in Fig. 5.5 (Bowman, 1977) and emphasize that agricultural systems may have to become more labour intensive as fossil fuel supplies decline and the costs of energy subsidies rise beyond the means of developing countries.

Jensen (1978) suggests that the ceiling to increased wheat productivity in the U.S.A. will occur before the turn of the century and that the maximum future increases in production will be about 27 per cent. The population of the U.S.A. will increase by 24 per cent over the same period and effectively eliminate that country's grain surpluses if current patterns of usage continue. In India the average grain supply *per capita* is about 180 kg per year, of which most is eaten. In the U.S.A. and Canada the average grain supply is 1000 kg, of which only 90 kg is eaten directly and the remainder fed to livestock (Brown and Eckholm, 1975). Sixty-nine per cent of the protein in these countries is provided by animal products. The logic is inescapable that high Western nutritional standards cannot be attained by most of the world and are unlikely to be maintained by these countries in the future. Allaby (1977) concludes from an extensive analysis of world food resources that adequate diets could be

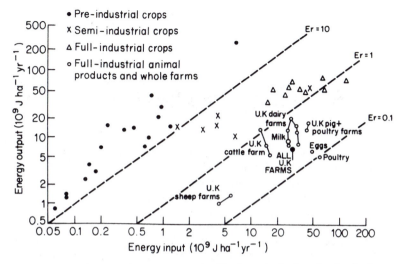

Fig 5.5 Energy inputs and outputs per unit of land area in food production in the world. Those points below the line Er = 1 represent systems in which the energy input is greater than the energy output, for instance the point for poultry represents a production system in which energy input is ten times the energy output. The most efficient production systems in energy terms (above the line Er = 10) are in the main peasant crop production systems in which the energy input is largely human rather than fossil energy. (Adapted from Bowman, 1977 after Leach, G. (1976). *Energy and Food Production*. IPC Science and Technology Press, Guildford.)

provided for the world population at the turn of the century and beyond, but this requires radical social and economic reforms. The populations of rich countries must consume less grain-fed livestock and the poorer countries improve animal protein production. A more equitable distribution of energy and technical resources is also required.

6 Secondary Production

Primary production involves essentially one trophic level, though there may be considerable variation in the production efficiencies of different plant species and communities. Secondary production in ecosystems involves not only a more physiologically diverse assemblage of organisms but also the summation of production within the various trophic levels of the herbivore and decomposer subsystems. The transfer of energy and nutrients through food-webs is consequently affected by a large number of steps. The ecological efficiencies of these transfers are determined both by thermodynamic and biochemical laws, and also be ecological parameters which vary in space and time. There is, therefore, no simple relationship between primary and secondary production at any level of ecological organization.

The empirical determination of secondary production, even in a single species population with a simple life history, is a complex ecological exercise. The basic approach is to sample a single cohort or generation at discrete time intervals and determine growth increments, natality and mortality between successive samples. The sum of growth and mortality is equivalent to the total production of the cohort and, where the species breeds only once a year, the cohort production is equal to annual production. More complex life histories necessarily involve more complex analysis. Details of the methodology for estimating the productivity of terrestrial animals are given in Petrusewicz and Macfadyen (1970); production by fish populations is covered by Bagenal (1978). Various short-cut methods of estimating secondary production are described by Phillipson (1970) and Waters and Crawford (1973).

Production estimates for a large number of animal species, mainly herbivores, are available in the literature. Few estimates, however, have been made of community production and no complete studies for major ecosystem types. Attention has therefore been directed, particularly under the auspices of the International Biological Programme (see p. 38) towards the formulation of models which will permit the prediction of secondary production both within trophic levels and between ecosystems. Detailed consideration of these models is beyond the scope of this account. Krebs (1978) provides a more comprehensive discussion of terrestrial and aquatic secondary production. The following section illustrates the basic principles of energy flux within and between populations and trophic levels.

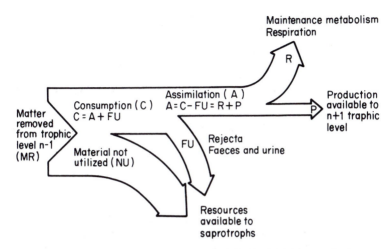

Fig. 6.1 Resource utilization by an animal of trophic level 'n'. The terminology follows Petrusewicz and Macfadyen (1970). For details see text.

Energy Flow within Populations

A general scheme for the partitioning of energy within an animal population is shown in Fig. 6.1. The material removed (MR) from a previous trophic level is rarely all utilized and some proportion of the tissues is wasted (NU). The amount of non-utilized material can be a significant proportion of MR – elephants or beavers may eat a small proportion of the trees they destroy, rodents may bury seeds which are subsequently not recovered and predators rarely consume whole prey. It is often difficult to differentiate between food ignored and food wasted and so consumption (C) is usually taken as a measure of gross energy intake from the previous trophic level. Some proportion of the ingested material is assimilated and the remainder egested as faeces (F). Urine (U) is included with the faeces (FU) because, although urine is a product of metabolism, it represents energy and nutrients which are not transferable through secondary production. Assimilation (A) for practical purposes is designated as the difference between consumption and rejection:

$$A = C - FU$$

The assimilation efficiency (A/C) is broadly related to diet (saprotrophs < microbivores ≈ herbivores < carnivores) i.e. it reflects the similarity in composition between food and consumer tissues. Average values for assimilation efficiencies, according to taxonomic and trophic identity, are given in Table 6.1. Considerable variation in assimilation efficiencies is also found within trophic levels, particularly for saprotrophs. Many

Table 6.1 A simple taxonomic-trophic categorization of heterotrophic organisms. For each category the characteristic assimilation (A/C) and growth (P/A) efficiencies are given. (From Heal and Maclean, 1975.)

	Herbivore A/C	Herbivore P/A	Carnivore A/C	Carnivore P/A	Microbivore A/C	Microbivore P/A	Saprotroph A/C	Saprotroph P/A
Microorganisms	—	—	—	—	—	—	—	0.40
Invertebrates	0.40	0.40	0.80	0.30	0.30	0.40	0.20	0.40
Vetebrate homeotherms	0.50	0.02	0.80	0.02	—	—	—	—
Vertebrate heterotherms	0.50	0.10	0.80	0.10	—	—	—	—

litter-feeding animals have assimilation efficiencies of 15–30 per cent. Nutrient availability, rather than energy, limits production by saprotrophs and nutrient uptake from low quality food necessitates the rapid processing of large volumes of material. There has been no selection for a high digestive efficiency of plant structural polysaccharides, particularly cellulose and lignin, which would result in the release of energy which the animals cannot use. Aphids which excrete large amounts of sugars are in a similar situation (p. 82). Snails and earthworms, however, do have cellulases (enzymes for breaking down cellulose) and assimilate 70–90 per cent of ingested material. This may reflect the carbon demand for the copious production of mucropolysaccharides by these animals. Most litter and wood-feeding termites have similarly high assimilation efficiencies but, like other social insects, have high respiratory rates. Additional energy may be needed to subsidize nitrogen fixation by symbiotic gut bacteria (Swift *et al.*, 1979). Most herbivores have assimilation efficiencies of 30–60 per cent but assimilation varies with food quality. Average values for ruminants feeding on natural vegetation are around 50 per cent. Carnivores and carrion feeders assimilate 70–90 per cent of ingested food.

The assimilated material is used for maintenance (basic metabolism and activity) and production (growth and reproduction); maintenance may be measured as metabolic heat or more usually as respiration (R)

$$A = R + P$$

Maintenance costs in animals involve nutrients as well as energy. In humans, for example, approximately 8 per cent of body protein is excreted daily (as nitrogen in urea) and resynthesized from assimilated amino acids. This is why malnutrition is not simply a question of protein deficiency *per se* but a lack of high quality protein, such as that found in animal products (i.e. a balanced complement of amino acids similar to that of human tissues). If a key amino acid is missing, as is the case with grains, a lower proportion of the total protein is metabolically useful. A combination of legumes and grains in a diet results in more usable protein than either food taken alone (Brown and Eckholm, 1975). Ruminants are able to excrete a proportion of the urea into the rumen where microbial symbionts convert it into utilizable protein. Under conditions where food quality is low the kidneys excrete less urea and nitrogen is conserved (Hobson, 1976).

Homoiotherms (warm-blooded animals) have a high metabolic cost of maintaining body temperatures and respire over 95 per cent of assimilated energy. The production efficiencies (P/A) for most homoiotherms therefore range from about 1.5 per cent to 3 per cent compared with a range of 20 to 40 per cent for most insects. These differences in production are generally attributed to the high cost of homoiothermy and are illustrated

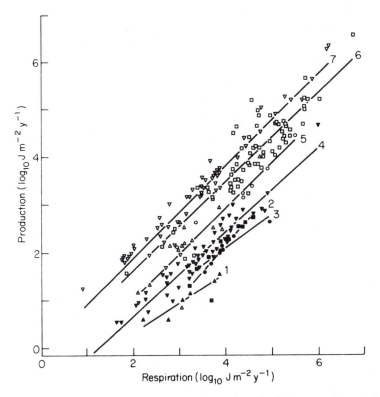

Fig. 6.2 The relationship between respiration and production in natural populations of animals. The regression lines for seven groups are shown: 1 = insectivores (▲), 2 = small mammals (■), 3 = birds (●), 4 = other mammals (▼), 5 = fish (○) and social insects (△), 6 = invertebrates other than insects (□), 7 = non social insects (▽). (After Humphreys, W. F. (1977). *J. Anim. Ecol.*, **48**, 427–53.)

in Fig. 6.2. In general terms the regressions of production on respiration all have a slope of 1 but the intercepts on the respiration axes move from around the origin for insect heterotherms (poikilotherms or cold-blooded animals) to a significant positive respiratory increment for homoiotherms. Fish and social insects occupy an intermediate position between the two major trends and this is reflected by production efficiencies for fish populations of about 10 per cent. The P/R relationship for insectivores shows a significant departure from that of other small (herbivorous) mammals. The surface area to volume ratio of a body increases linearly with decreasing body size and therefore small homoiotherms have disproportionately higher metabolic rates than larger animals in order to compensate for more rapid body heat losses. Predators also tend to have higher respiratory costs of food capture than herbivores and so small homoiothermic predators such as shrews may respire 99 per cent of assimilated energy. Humming birds and insectivorous bats are able to relax physiological control over temperature regulation and allow body temperatures to track ambient temperatures when resting. The energy savings by humming birds may amount to 90 per cent of the energy required to maintain normal operational body temperature (Hainsworth and Wolf, 1970).

The production efficiencies of animals tend to vary with food quality, the length of the life cycle, age structure (the production efficiency of many young herbivorous homoiotherms is about 35 per cent compared with 3 per cent for the adults) and, in the case of heterotherms, with environmental temperatures. Some average values for various groups of organisms within different trophic levels are given in Table 6.1.

Production and Energy Flow between Trophic Levels

The next level of regulation of production comes from the organization of species populations into the trophic levels of the herbivore and decomposer subsystems (Fig. 6.3). Several important conceptual differences distinguish the two trophic systems (Heal and MacLean, 1975). Firstly, the herbivore system is biotrophic at its base (i.e. the $1°$ consumers utilize living plants) and events occurring within the system can interact with and modify the rate of energy input as NPP, as in overgrazed pastures. The decomposer system is saprotrophic at its base (the $1°$ consumers utilize dead materials) with no direct effect upon the rate at which energy enters the system; although decomposition may indirectly affect NPP.

Secondly, energy passing through the herbivore subsystem may be respired or pass to the decomposer subsystem whereas energy entering the decomposer subsystem is only lost through respiration (assuming an equilibrium system where P:R $\approx$ 1). Heal and MacLean (1975) conclude that the decomposer subsystem is conservative of energy through recycling (see Fig. 6.3) and that this feature, together with the major proportion of NPP which enters the system, is critical in allowing the

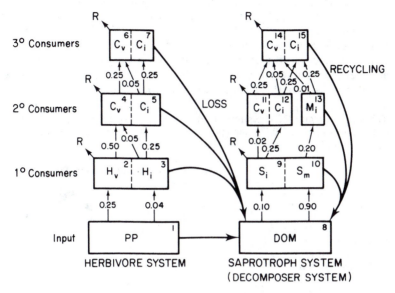

Fig. 6.3 A generalized trophic structure for terrestrial ecosystems and the consumption efficiencies ($C_n/P_n - 1$) used in the calculation of heterotrophic productivity in grasslands. PP = primary production and DOM = dead organic matter. For further abbreviations see Table 6.2. (From Heal and MacLean, 1975.)

greater length, complexity and biomass of decomposer than herbivore food chains.

The key parameter governing the flow of energy through the two systems is the consumption efficiency ($C_n/P_n - 1$) which is a measure of the consumption by trophic level n of production by the previous trophic level, $n - 1$. Some generalized values for consumption efficiencies are shown in Fig. 6.3. It is fairly well established that herbivourous insects usually consume less than 5 per cent of NPP, with vertebrate herbivores consuming a maximum of 20 per cent in grasslands. In temperate regions saprotrophic invertebrate animals consume around 10 per cent of litter fall and the remaining 90 per cent is utilized by microorganisms. In tropical savannahs the direct contribution of termites to decomposition processes may be much higher and most woody litter may be processed by the Macrotermitinae. Consumption efficiencies for predator/prey links are more speculative. Vertebrate predators may exceptionally predate 100 per cent of vertebrate production though a value of 50 per cent was considered by Heal and MacLean (1975) to be more usual. Vertebrate predators rarely consume more than 5 per cent of invertebrate prey while invertebrate predator/prey consumption efficiencies are estimated at 25 per cent.

The assimilation and production efficiencies shown in Table 6.1 were used in the trophic level model to calculate heterotroph production in ecosystems where values of NPP and heterotroph production were known. Consumption efficiencies for vertebrate herbivores were taken as 5 per cent for non-grassland sites. The predicted values were generally within an order of magnitude of observed levels of production. Calculated rates of energy flow through a grassland ecosystem are shown in Table 6.2.

Table 6.2 Calculated ingestion, production, respiration and egestion by heterotrophs (kJ m^{-2} yr^{-1}) per 100 kJ m^{-2}* net annual primary production in a grassland ecosystem. (After Heal and MacLean, 1975.)

	Ingestion	Production	Respiration	Egestion
HERBIVORE SYSTEM				
Herbivore				
vertebrate (H_v)	25.000	0.250	12.250	12.500
invertebrate (H_i)	4.000	0.640	0.960	2.400
Carnivores				
vertebrate (C_v)	0.160	0.003	0.123	0.031
invertebrate (C_i)	0.170	0.040	0.095	0.034
SAPROTROPH (DECOMPOSER) SYSTEM				
Saprotrophs				
invertebrate (S_i)	15.153	1.212	1.818	12.122
microbial (S_m)	136.377	54.551	81.826	—
Microbivores				
invertebrate (M_i)	10.910	1.309	1.964	7.637
Carnivores				
vertebrate	0.041	0.001	0.032	0.008
invertebrate	0.648	0.155	0.363	0.130
TOTAL	192	58	99	35
% passing through				
Herbivore system	15.2	1.6	13.5	42.9
Saprotroph system	84.8	98.4	86.5	57.1

*The original figures were in kcals and only the units have been changed since values are expressed on a proportional basis.

It is consistent with most terrestrial ecosystems that the herbivore subsystem only accounted for 1.6 per cent of heterotroph production and that 94 per cent of total production was by fungi and bacteria. Consequently the differing consumption efficiencies for vertebrate herbivores in grassland and forest ecosystems have little effect on total patterns of heterotroph productivity. These figures also emphasize that the majority of secondary production in ecosystems is in the form of microbial tissues which are not utilizable by man. Heal and MacLean (1975) concluded that the model provides a means of predicting the production by various taxonomic–trophic categories in ecosystems and of exploring the effects of variation in the composition and efficiencies of different groups of organisms. By adjusting the input from NPP to correspond to regional variations approximate scales of heterotrophic production can be obtained on a global scale.

Phillipson (1973) has carried out a similar exercise in estimating protein production for the Serengeti Plains ecosystem from production:biomass ratios (Fig. 6.4). The model predicted the annual production of 0.2 $g\,m^{-2}$ (2 kg ha^{-2}) utilizable protein by large mammals. The production of wild herbivores and beef cattle was then compared by taking the maximum carrying capacity of European-managed ranches in East Africa of 1.32 g m^{-2} dry wt, applying a production to biomass efficiency of 11.5 per cent

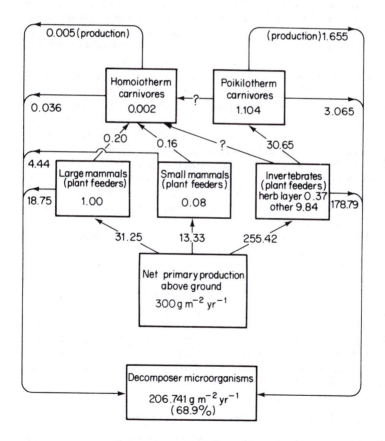

Fig. 6.4 A tentative model of the Serengeti Plains (East Africa) ecosystem. Except where indicated the figures in the boxes are for biomass (g m^{-2}), all other figures are annual fluxes (g m^{-2} yr^{-1}). (From Phillipson, 1973.)

for beef cattle. The calculated production of 0.15 g m^{-2} yr^{-1} (dry wt) suggests that production by beef cattle on the best ranches in East Africa is only 75 per cent of the value attained by wild ungulates on the same type of pasture. Experiments with ranching wild antelope have confirmed that many species, particularly eland and orynx are not only more productive and disease resistant than cattle in the drier regions of Africa but also less damaging to natural vegetation (Jaffe, 1975).

A final example of secondary production estimates using ecological efficiencies is Ryther's (1969) analysis of the potential productivity of the oceans. The oceans were divided into three regions with different levels of primary productivity — the open ocean, coastal zone and coastal upwelling areas (Table 6.3). Fish production was calculated using Lindemann's transfer efficiency (A_n/A_{n-1}) with values of 10–20 per cent between each trophic level. Fish production in the fertile upwelling zone was equal to that of coastal regions with nearly 100 times larger surface area. Ryther comments that 'The open sea, nearly 90 per cent of the ocean and three quarters of the world's surface, is essentially a biological desert. It produces a negligible fraction of the world's fish catch and has little or no potential for yielding more in the future.' Of the 242 million tonnes of fish, Ryther estimated that approximately 100 million tonnes could be harvested. These figures have been challenged on the basis of the number of trophic levels, the value of transfer efficiencies and the sustained yield of fish populations, and both higher and lower limits have been suggested. Fish catches, however, which have risen by 5 or 6 per cent per annum since 1950, appear to have peaked around 70 million tonnes, despite the fact that there are several fish species which have yet to be commercially exploited. Energy costs and availability may impose restraints on further increases in fish catches.

The decline in world fish catches by 4.7 million tonnes in 1972 was associated almost entirely with the collapse of the Peruvian anchovy fishery. From 1968 to 1971 Peru was the world's leading fishing nation and accounted for more than 16 per cent of the entire world catch. In 1971 anchovy landings were 13 million tonnes but by 1973 they had crashed to only 2 million tonnes. A combination of excessive over-fishing and temporary changes in the patterns of coastal currents, which cause the upwelling of nutrient-rich waters, are blamed for the decline. The anchovy fisheries have shown little subsequent recovery and join the list of over-exploited fish stocks of the world which include haddock, hake, herring, cod, plaice, tuna, mackerel and sardines.

Livestock Production and Other Sources of Protein

The varieties of livestock which have been domesticated are more limited than for plants. Cattle, pigs, sheep, goats, water buffalo, turkeys, geese, ducks and chickens account for nearly 100 per cent of protein production

Table 6.3 Estimates of primary production and fish production in three zones of the ocean. (From Ryther, 1969.)

Zone	% of ocean	Mean primary productivity (gC m^{-2} yr^{-1})	Total primary production (10^9 t C yr^{-1})	Number of trophic levels	Efficiency of energy transfer	Fish production (t wet weight)
Open ocean	90	50	16.3	5	10	0.16 × 10^7
Coastal zone*	9.9	100	3.6	3	15	12.0 × 10^7
Upwelling areas	0.1	300	0.1	1 or 2	20	12.0 × 10^7
Total			20.0			24.16 × 10^7

*Includes some offshore areas of high productivity.

by domestic animals. Cows produce more than 90 per cent of milk consumed, water buffalos about 4 per cent and goats and sheep most of the remainder (Ehrlich *et al.,* 1977). World production of meat in 1973 amounted to approximately 108 million tonnes (comprising 38 per cent pigs, 37 per cent cattle and buffalo, 18.5 per cent poultry, 6 per cent sheep and goats and about 0.5 per cent horses) of which over two-thirds was produced in Europe and North America. In temperate regions, where the pasture is good, cattle are mainly grass-fed but in North America and cold climates, where the demand for meat or dairy products exceeds the carrying capacity of the pasture, they are largely fed on grain. Pigs and poultry are not only fed grain but also animal protein concentrates (principally fish meal). The conversion efficiencies for protein are low. Broiler chickens convert only about 18 per cent of feed protein into utilizable meat for human consumption, pigs about 9 per cent and bullocks (steers) 6 per cent. Put another way 1 kg of chicken, pork or beef is equivalent to 5.4 kg, 10.6 kg and 15.8 kg of grain protein respectively (Pimentel and Pimentel, 1979). Intensive grain feeding of livestock started in the 1950s and 1960s when cereals were cheap and abundant. During the period 1961 to 1973 world production of milk increased by 21 per cent, meat by 42 per cent and eggs by 43 per cent. This increase largely occurred in the developed countries where the indirect consumption of grain as animal products increases as a function of national affluence (Fig. 6.5).

There are many hidden energy costs in meat production, over and above those involved in grain production, which make meat an expensive commodity in terms of energy resources (see Fig. 5.5). The production efficiency of farm livestock can be increased by raising food quality (i.e. assimilation efficiency) and reducing the metabolic cost of activity and basic metabolism. Factory farms reduce metabolic costs by restricting the movements of animals and provide an energy subsidy for the animals by heating the buildings.

The main ways in which the efficiency of livestock populations can be increased, rather than changing the production efficiency of individual animals, are discussed by Holmes (1977) and Bowman (1977) and include the following.

(i) The proportion of males which have to be maintained to mate females. Artificial semination significantly reduces the ratio of males to females and increases population efficiency.

(ii) The reproductive rate of females. This includes both the proportion of female offspring which are fertile and their fecundity.

(iii) Mortality at all stages of the reproductive cycle.

(iv) The age at which parents have to be replaced. This factor is affected by mortality, the longevity of the parents (reproductive life) and the replacement rate necessary to improve the genetic potential of the stock.

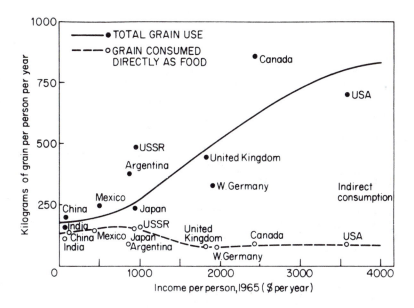

Fig. 6.5 Direct and indirect grain consumption. As incomes rise, per capita grain consumption rises rapidly. Direct grain consumption tends to be lower in the rich than poor countries but the amount of grain indirectly in the form of meat, milk and eggs is several times higher. (From *Food Balance Sheets. 1964/66 Average*. U.N. Food and Agriculture Organization, Rome (1971).)

In Fig. 6.6 the population efficiency output of the different farm species is given in relation to different levels of reproduction or output. These efficiencies take account of primary output alone and do not include the energy content of faeces and urine. At high levels of reproduction, little increase in production efficiency can be attained by further increases in reproductive rate. However, increases in the output of milk per animal substantially increases the efficiency of dairy cattle (Bowman, 1977).

It is not surprising that, historically, selection of domestic stock has not been based on ecological principles and reappraisal of animal protein production methods may be required in the future (Holmes, 1977). Small rapidly growing animals are more productive than large animals with an extended period of growth. Phillipson (1966) points out that a bullock produces a 485 kg carcase from 7400 kg of feed in 120 days. The same amount of meat is produced by 300 rabbits from the same amount of food in 30 days. The possible animal protein yield from the produce of one hectare of agricultural land is shown in Table 6.4. This emphasizes the point made earlier (p. 114) that in the long term the consumption of animal protein by developed countries must include a larger proportion of dairy products.

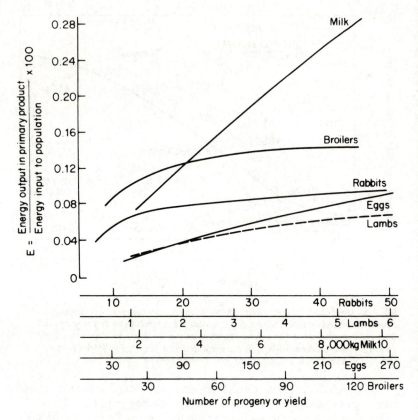

Fig. 6.6 Population efficiency of different species. The efficiency of a species depends on the production level and changes to a greater extent for dairy cattle and laying hens than for meat animals. (From Bowman, 1977; after Spedding, C.R.W. *The Supply of Food*. Paper presented at 'Man and His Environment', Symposium at the University of Birmingham, September, 1975.)

Aquaculture has been carried out using a wide variety of bivalve molluscs (mussels and oysters) and crustacea (shrimps, crabs and lobsters) but fish farming, particularly in Asia, is one of the most intensive methods of protein production. The World Health Organization have estimated that fish yields from inland waters, including at least half from aquaculture, were in the order of 17 million tonnes in 1980 with a potential of 50 million tonnes by the turn of the century. The yields of dry protein can be as high as 675 kg ha^{-1} in comparison with 78 kg ha^{-1} (dry weight) beef protein from intensive grazing but levels of production are to a large extent, as with crop production, proportional to nutrient input and culture technology.

Table 6.4 The yield of product, edible protein and edible energy which can be obtained from breeding populations fed from the produce of 1 ha. (From Holmes, 1977.)

	Product (kg)			Edible protein (kg)	Edible energy (MJ)
	Carcass	Eggs	Milk		
Eggs	85	1250		138	8900
Broiler	1225[a]			137	7500
Turkey	1000[b]			144	4200
Rabbit	730			118	4800
Bacon					
12 piglets per year	745			80	7700
24 piglets per year	900			98	9300
Sheep					
1.4 lambs per year	268			32	3500
2.8 lambs per year	423			50	5500
Suckler cow					
0.9 calves per year	255			35	2800
1.8 calves per year	365			50	4100
Milk with low concentrates[c]	60		3940	138	11500
Milk and 1.8 m beef[d]	166		2800	116	9500
Milk and 24 m beef[d]	162		2600	110	9400
Milk with high concentrates[e]	52		4100	142	11790
Milk and veal[d]	100		3900	144	11500
Milk and cereal beef[d]	1355		3100	120	9700
Milk and 18 m beef	150		3000	120	9900

[a]Skin is included. If excluded, values are reduced to about 70% of values shown.
[b]Skin is included. If excluded, values are reduced to about 80% of values shown.
[c]Low concetrates, 900 kg per cow per year.
[d]The calves surplus to those needed to replace the milking herd are reared for veal, cereal beef, etc.
[e]High concentrates, 1650 kg per cow per year.
Sources: Holmes, W. (1974). Chap. 58 in *Meat* (D.J.A. Cole and R.A. Lawrie (eds)), Butterworth, London. Holmes, W. (1975). The livestock in Great Britain as food producers. *Nutrition, Lond.*, **29**, 331–6.

On a global scale, most of the feed used in aquaculture is of plant origin (rice bran, soy bean meal, peanut meal and oil-cake are commonly added to fish ponds in Israel, India and S.E. Asia) as a consequence of the increasing cost of even low-grade protein. Recently experiments have successfully been carried out in the U.K. on the use of processed human faeces (which are rich in fats, protein and vitamins) as feed supplements for pigs and poultry as well as fish. In Israel very high rates of fish production have been achieved by replacing conventional pelleted feeds with cow manure and it is estimated that 10 tonnes dry manure is converted to 4 tonnes live weight of fish (Reay, 1979). Fish culture in sewage lagoons has been practised in China and S.E. Asia for centuries.

Temperature is probably one of the most important environmental parameters affecting the growth rates of animals in aquaculture but heating the ponds in temperate regions greatly increases the cost of the products. However, the effluent water from some power stations in North America, U.S.S.R., Europe and Japan is typically $8-10°C$ above ambient temperature and is used in fish farms. Reay (1979) cites the example of Ratcliffe on Soar (U.K.) where carp held in condenser water (mean temperature $25°C$) grew to 970 g in 10 months compared with only 180 g over the same period in river water at a mean ambient temperature of $15.5°C$. Also lobsters, which take about six years to reach 450 g in the wild, reached the same weight in heated effluent at $22°C$ in 2 years.

The yields are to some extent, as with crop production, proportional to culture technology and fertilizer input. Aquaculture thus not only offers a means of producing a high quality animal protein from a limited land area but also a means of recycling nutrients in human and livestock effluent. In addition, there are some 440 million hectares of coastal wetlands in the world (mostly mangrove swamps) and if only 10 per cent of this area could be used for aquaculture at least 100 million tonnes per annum of extra fish could be produced (Reay, 1979).

A novel protein source which is receiving increased attention is fungal growth on wheat and rice straw. The bulk of straw waste is burned, causing not only an environmental nuisance but also a loss of organic nitrogen. Fungal production on straw can be highly efficient. The common cultivated mushroom, *Agaricus bisporus*, can convert 100 kg of fresh wheat straw into 50 kg of fruiting bodies (Hayes and Lim, 1980) though up to 95 per cent of the fresh weight is water. Over and above the value of the fruiting bodies for human consumption the spent straw can be a valuable feed for ruminants. Fungal mycelium has a similar nutritional value as other conventional protein supplements for livestock feeds (Zadrazil, 1980). Cellulose and lignin are also converted into more digestible energy sources for monogastric (i.e. non-ruminant) animals.

World production of the two principal mushrooms *Agaricus bisporus* (in temperate areas) and *Volvariella volvacea* (in the tropics) amounted to 420 000 tonnes in 1975 (Declaire, 1978). While this is a minute proportion of the current protein demand of over 40 million tonnes, mushrooms have great potential for the conversion of waste straw into protein on a local scale. (Due to its bulk the cost of straw transport and storage is a major economic factor in the commercial production of mushrooms.) The by-products of a wide range of industrial processes, such as molasses, tree bark and municipal refuse also have potential as substrates for fungal growth.

Current investigation of the value of fungi as protein sources is being concentrated not on mushrooms but on microfungi, particularly yeasts (Cooke, 1977) as the protein content of yeasts can exceed 40 per cent and the unicellular growth form, together with high growth rates, is appropriate for industrial processes. Strains of *Torula* and *Candida* are the principal yeasts used industrially and these can be grown on a variety of substrates including waste products from sugar refining, whey from milk processing and even wood-pulp sulphite liquors from paper-making (which otherwise create serious problems when discharged as effluent into rivers). Commercial plants are also coming into production which grow yeasts on n-alkanes (middle distillates from petrol refining) and methanol. The projected annual yield from one plant in the U.K. is 100 000 tonnes of yeast which can be used as a substitute for milk, fish and soya protein at dietary levels of about 30 per cent in pig and broiler fowl feeds (Cooke, 1977). Single cell protein foods for humans are likely in the future but will probably be more expensive than animal feed stuffs to produce.

7 Decomposition

The same basic principles underly energy and nutrient transfers within the herbivore and decomposer subsystems but the overall trophic dynamics are fundamentally different. Organic matter entering the decomposer subsystem is transferred through the various trophic levels but non-utilized material, rejecta and secondary production form the inputs to further cycles. Thus with each passage of organic matter through the organism, complex energy, carbon and soluble mineral nutrients are released until mineralization is theoretically complete. In the model considered earlier (Fig. 6.3), 57 per cent of the energy entering as dead organic matter was lost as respiration with each passage and 43 per cent was recycled as organic matter. Recycled material fell below 1 per cent of NPP after 6 cycles $(43\% - 18.5\% - 8.0\% - 3.4\% - 1.5\% - 0.43\%)$. This model is conceptually useful for understanding the gross functioning of decomposer communities, it is not a model of decomposition *per se*. Litter decomposition does not conform to a simple mathematical series of weight losses. With each recycling a larger proportion of decay-resistant fractions of the litter remain and even more recalcitrant complexes are formed so that the rate of decomposition decreases with time. A negative exponential decay curve is the simplest theoretical model of decomposition but observed weight losses usually show a more complex pattern than this, according to the *composition of the resource* and the *characteristics of the decomposer community*. *Physical environmental parameters* (climate as well as local variables such as pH, oxygen status and temperature/moisture regimes) will further shape the characteristics of decomposition.

The Regulatory Variables of Decomposition Processes

A descriptive model of decomposition should include parameters for the organisms (O), the quality of the resources (Q) and the physical environment (P) because it is a combination of these regulatory variables which dictates the rate of change of each step in the decay process. This is represented by the simple decomposition module shown in Fig. 7.1(a). The changes in the nature of the resource involves a variety of state changes such as fragmentation, enzyme action and the loss of water soluble materials. Decomposition (D) is thus a sum of the component losses attributable to catabolism (K), comminution (C) and leaching (L) (Swift *et al.,* 1979):

$$D = KCL$$

Catabolism is the enzymatic degradation of a compound from a polymer to a monomer (such as cellulose to glucose) or to its mineral constituents (glucose to CO_2 and H_2O).

Comminution is a reduction of particle size and differs from catabolism in being a predominantly physical rather than chemical process. It is largely brought about by animal feeding activities but abiotic processes (wind, freeze/thaw and wet/dry cycles) may also be involved.

Leaching is the removal of soluble materials by water and is an entirely physical process, though the rate of leachate losses is influenced by catabolism and comminution.

The decomposition of any organic material will involve these three component processes though any one may predominate under particular circumstances, for example P in deserts or tundra where decomposition is limited by moisture or temperature, O where termites are present or where resource quality (Q) limits decomposition.

Decomposition processes can therefore be visualized as a cascade of steps, each regulated by the complex of OPQ variables, with the products of the KCL processes forming inputs to separate modules as shown in Fig. 7.1(b). Physical and biochemical complexity will increase with time from the onset of decay and the organisms will show a corresponding increase in species diversity, some of which will be responsible for specific processes, others are non-specific. As mineralization progresses the rate of decomposition and diversity of resource types decreases until, theoretically, only the constituent minerals remain. The PQ variables determine the time scale and degree to which these processes are complete. In undisturbed organic soils this time sequence is represented in time by the different horizons of the soil profile, in the same way that a hydrosere represents a successional sequence. Organic matter moving down the profile is processed by new organism complexes under different environmental conditions, and structural complexity of the organic matter decreases with increasing depth. The accompanying changes in resource quality are reflected by decreasing microbial respiratory rates from the surface litter layers to the mineral horizons (Table 7.1). A spatial sequence of successive steps in decomposition processes is not usually observed in aquatic systems.

The different scales at which the regulatory variables operate are indicated in Table 7.2 and will be illustrated by the following discussion of organisms, resource quality and physical environment.

The Organisms

Organisms of decomposer food webs can be notionally classified into

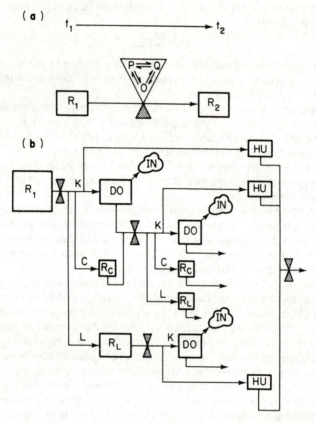

Fig. 7.1 Resource model of the decomposition subsystem and the driving variables.
(a) The basic decomposition module illustrating the regulation of changes in resource state (R_1 to R_2) over a short period of time (t_1 to t_2) by the driving variables: the organisms (O), the physico-chemical environment (P) and resource quality (Q).
(b) A simplified model of the cascade processes whereby a resource is dissociated by the processes of catabolism (K), comminution (C) and leaching (L) and the substrate components are mineralized (IN), or resynthesized into tissues (DO) and humus (HU). Soluble materials (R_L) may be removed in unchanged form to other sites. Different rates of change may occur during each stage and the balance of the regulatory factors may also differ. (After Swift *et al.*, 1979.)

necrotrophs, biotrophs and saprotrophs according to whether they kill their food resources, exploit living hosts or utilize dead materials. The suffix -troph is used in preference to -vores, since it does not imply food ingestion and can be used for animal and microbial trophic groups (Swift *et al.*, 1979).

The following section is concerned with the functional biology of saprotrophs. Detailed biology of the various taxonomic groups can be

Table 7.1 Rates of oxygen uptake by different horizons in a podzol under *Pinus sylvestris*. (From Parkinson, D. and Coups, E. (1963). pp. 167–75 in *Soil Organisms* (J. Doebsen and J. van der Drift, eds). North Holland, Amsterdam.)

Horizon	Characteristics	% organic matter	Oxygen uptake (μl O_2 5 hr^{-1})	
			Per gram dry soil	Per gram organic matter
OO (L)	Litter	98.5	2366.0	2406.0
O_1 (F_1)	Fermentation layer	98.1	1400.0	1428.0
O_2 (F_2)	Fermentation layer	89.3	245.2	274.6
O_2 (H)	Humus	54.6	80.9	148.3
A_1	Eluvial mineral horizon	17.2	13.3	77.7
A_2	(leaching of colloids and ions)	1.9	4.5	238.8
B_1	Illuvial horizon	10.6	9.8	91.9
B_2	(deposition of colloids and ions)	5.2	2.9	56.6
C	Mineral horizon over bedrock	1.4	1.4	96.3

found in general texts on soil biology (Burges and Raw, 1967; Dickinson and Pugh, 1974).

Fungi and bacteria

The function of fungi and bacteria as decomposers can be primarily assigned to successful adaptations in growth form and nutrition.

Growth form

Two major types of growth form are seen in microorganisms: the colonial growth form arising by divisions of unicells (yeasts and bacteria) and the filamentous growth form typical of most fungi and the Actinomycetales (a group of filamentous bacteria).

The filamentous growth form is of great significance in decomposition because of the ability it confers for the penetration and invasion of organic matter. Many fungi can produce boring hyphae which mechanically penetrate intractable portions of resources such as cuticularized leaf surfaces, seeds and pollen grains. Penetration is followed by the production of a frond-like mycelium which is able to exploit planes of weakness between cell walls and within molecular lattices of structural polymers. In cellulose, for example, strong bonds exist between molecules within the same chain which can only be enzymatically attacked, but only relatively weak hydrogen bonds link the molecules of adjacent side chains and these can be separated physically by the apical growth of fungal hyphae as well as by enzymic activity.

Table 7.2 The regulation of decomposition rate by organisms, physical environment and resource quality at different scales of operation. (From Swift et al., 1979.)

Scale of resolution	Resource type	Organisms	Scale of operation environment	Resource quality
Ecosystem	Total detritus	Decomposer community	Macroclimate Edaphic complex	Relative proportions of main resource types
Ecosystem component	Main resource types (primary, secondary etc.) down to taxonomic origins of components (leaf species etc.)	Resource specific selection within community	Local variation in environment (down to level of resource size, i.e. microenvironment)	Chemical and physical composition of resources
Resource components (substrates)	Cellulose, lignin, keratin, etc.	Enzymatic capability of species	Microenvironment down to molecular environment of enzymes, etc.	Substrate specificity of enzyme systems

The formation of an extensive mycelium is also an important adaptation for microbial growth in a physically heterogeneous habitat such as the soil. The mycelium allows the translocation of nutrients between microsites so that resources which are potentially limiting for fungal metabolism in one microsite can be supplied from elsewhere in the mycelial complex. Translocation usually occurs over a few microns but some wood decaying fungi (Basidiomycetes) form mycelial strands, or rhizomorphs, which can translocate materials over several metres.

Whereas filamentous microorganisms are adapted to penetration and exploitation of relatively massive materials over long time scales, the unicellular microorganisms are adapted to surface microhabitats which are rapidly exploited in the short-term. The unicellular growth form does not facilitate penetration of materials but the small size of the bacterial cell does facilitate the colonization of fine cavities and pore spaces. Bacteria are thus adapted to the utilization of particulate material where the surface area to volume ratio is high.

Bacterial and yeast colonies grow by cellular fission and the daughter cells are separate entities. Hyphal growth occurs by apical elongation and, in many fungi, cross walls or septa occur at infrequent intervals along the hyphae and delimit multinucleate coenocytes. The bacteria and yeasts are therefore pre-adapted for growth under disturbed conditions while fungal growth can be affected by disruption of the mycelium. Actinomycetes are between these extremes and the filamentous thallus can fragment and function as unicells.

Microorganisms have a limited capacity to spread, except through spore dispersal, but the vegetative structures of fungi and bactera show a wide range of methods for resisting unfavourable conditions. These include the production of resistant spores which can remain dormant for long periods of time. Bacterial endospores can survive extreme conditions of temperature and dessication, such as those found in deserts, for many years and are also resistant to high concentrations of metal ions (Gray and Williams, 1971). Most microorganisms can reproduce sexually (which maintains genetic diversity) and by rapid asexual reproduction which allows the exploitation of ephemeral resources. The various cycles of exploitation, reproduction, dispersal (dormancy), colonization and exploitation, etc. are fundamental in microbial ecology and many taxonomic groups have developed different strategies which partly explain their successful roles in decomposition processes (see Gray and Williams, 1971).

Nutrition
Saprotrophic microorganisms function by excreting extracellular enzymes which break down polymers into simple molecules which are then absorbed. In contrast to animals there are insignificant metabolic costs, efficiencies or adaptations associated with ingestion. Collectively fungi and bacteria can utilize most natural organic compounds and many man-made materials as well.

The fungi are generally regarded as the main decomposers of plant materials in both terrestrial and aquatic habitats because it is within this group that there is the most widespread distribution of lignin and cellulose degrading enzymes. Only a limited number of bacteria possess these capacities but they are probably the main agents of these processes in extreme environments such as anaerobic microsites. Although there are some differences between fungi and bacteria in their nutrition, primarily in the cellulolytic and lignolytic abilities of Basidiomycete fungi, both groups can use most other metabolites. The main distinction between the two groups in terms of their relative roles in decomposition processes lies in the wider range of conditions under which bacteria can operate as compared to those of fungi. Thus bacteria are the dominant saprotrophic microorganisms under extreme temperature conditions and in anaerobic habitats, including those of animal guts.

The fauna

The simplest functional division of decomposer communities is by thallus or body width, into the microflora (fungi and bacteria) and the micro- and meso- and macrofauna (Fig. 7.2).

The *microfauna* under this classification is composed of the Protozoa, nematodes, rotifers, tardigrades, with the smallest Collembola and Acari included at the upper limits. Functionally this is a coherent group as none of these animals is involved in litter comminution. Included in the *mesofauna* are the Collembola (springtails) and other apterygotes, the Acari (mites), enchytraeid worms, most Diptera (fly) larvae and some of the genera of smaller Coleoptera (beetles). Most members of this group can attack intact plant litter but their sum contribution to litter breakdown is generally insignificant. Their major role in decomposition processes is in regulating microbial populations and reworking the faeces of the macrofauna. The termites are an exception and are probably the only members of the mesofauna whose presence or absence can directly determine energy and nutrient flux pathways in terrestrial ecosystems. The *macrofauna* consists of the large litter-feeding arthropods such as the Diplopoda (millipedes), Isopoda (woodlice), Amphipoda and insects, as well as the molluscs (slugs and snails) and the larger earthworms. These animals are responsible for the initial shredding of plant remains and its redistribution within decomposer habitats. Their presence can significantly affect decomposition pathways and contributes directly to the structure of the soil.

The roles of these different groups of organisms in litter breakdown can be investigated using mesh bags. The results of a typical litterbag study are shown in Fig. 7.3 (Anderson, 1973). Two species of leaf litter, chestnut (*Castanea sativa*) and beech (*Fagus sylvatica*), were placed in two contrasting woodland soil types in South East England. The litterbag mesh sizes were chosen with some prior knowledge of the structure of these communities.

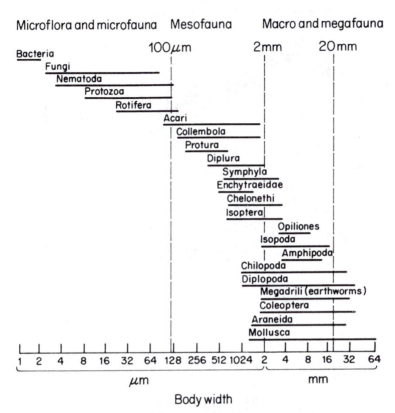

Fig. 7.2 Size classification, by body width, of organisms in decomposer food webs. (From Swift *et al.*, 1979.)

The largest mesh size allowed access to the entire soil and litter organism community and was theoretically a measure of total weight losses attributable to leaching, microbial catabolism and losses of leaf fragments less than 5 mm diameter following comminution. The medium mesh excluded the macrofauna but allowed entry to the meso- and microfauna, while the finest mesh excluded all organisms except the microflora, microfauna and the smallest members of the mesofauna.

Weight losses from the fine mesh bags were similar for each species in the two sites but consistently greater from chestnut than beech. This could be taken as a measure of microbial catabolism but weight losses from similar bags suspended above the soil, where microbial activity would be reduced by dessication, revealed that a significant fraction of the weight losses from the fine mesh bags were attributable to leaching. The rapid weight losses from litter over the first few weeks in the

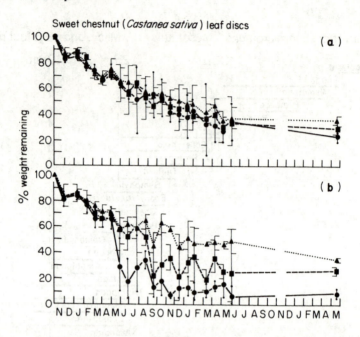

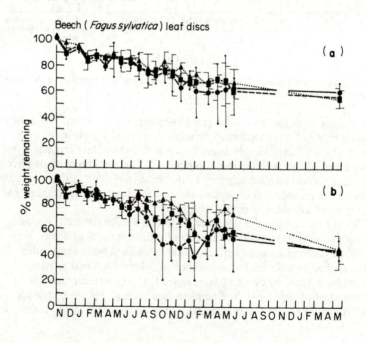

field (Fig. 7.3) were mainly attributable to losses of soluble materials.

The *Castanea* site has a mor-like humus form where layers of acid, organic matter accumulate on top of the mineral soil (see Table 7.1). The fauna, as in most organic woodland soils, was dominated by mesofauna while the macrofauna were scarce or absent. The negligible weight loss differences between the three mesh sizes for both litter types indicate that decomposition was dominated by microbial catabolism and leaching, and that litter comminution by the mesofauna was negligible. In fact the large chestnut leaves were largely broken down by wind action after they had been weakened by microbial decay.

In the *Fagus* site litter breakdown was rather different. The soil had a higher pH and mull-like characteristics where the organic and mineral horizons have been mixed by the activities of the macrofauna, particularly burrowing earthworms. Weight losses from the coarse mesh bags were higher than those in the *Castanea* site, due to the feeding activities of worms and millipedes, and only small amounts of litter remained in the bags after a few months in the field. Higher weight losses from the medium mesh bags were attributable to juvenile earthworms and not to larger mesofauna populations.

This experiment illustrates a number of points concerning the functioning of decomposer communities.

(1) The effects of resource quality can be seen in the different decay rates of the two leaf species (though chemical analysis failed to determine the basis for this difference).

(2) The complex of soil type and associated animal community affects the pattern of litter decomposition.

(3) The weight losses from litter bags are a comparitive but not an absolute measure of decomposition processes in soils. This is evident from the fact that while weight losses from the chestnut leaves were twice those from beech the level of soil organic matter accumulation in the *Castanea* site was twice that of the *Fagus* site. As comminuted material moves down the soil profile, or is covered by further litter falls, the decomposer environment, the organisms and decay rates change. Unless the litter bags follow this natural progression, which may or may not occur according to the soil type, errors are introduced into longer term decay estimates.

Fig. 7.3 Decomposition of leaf litter in two woodland soils with contrasting humus forms: **(a)** mor-like under sweet chestnut (*Castanea sativa*); **(b)** mull-like under beech (*Fagus sylvatica*). The soil fauna activities on the *Castanea* site are dominated by the micro- and meso-fauna while earthworms and macro-arthropods are important fauna elements in the *Fagus* site. Mean percentages ± 95 per cent limits (detransformed arcsin) of initial weights are shown for coarse, 5 mm (●), medium, 1 mm (■) and fine, 45 µm (▲) mesh litter bags containing chestnut and beech leaf discs. The mean and confidence limits for the final set of samples in May are not comparable with the other data because of larger sample sizes. (From Anderson, 1973.)

Similar methods and approaches can be used to analyse litter comminution rates in aquatic systems. Cummins (1973) proposed the following classification of freshwater invertebrates, based on their feeding habits.
(1) Shredders which bite large fragments from leaf litter.
(2) Gatherers and scrapers which collect food particles from the substratum and process the faeces of shredders.
(3) Deposit feeders which live in the bottom sediments and process fine detritus.
(4) Suspension feeders which filter suspended particles from the water.
(5) Grazers feeding on living plants.
(6) Predators.
The functional relationships between some of these groups are shown in Fig. 7.4. The roles of terrestrial and aquatic organisms in litter decomposition and nutrient cycling processes are considered in detail by Swift *et al.*, (1979) and in Anderson and Macfadyen (1976).

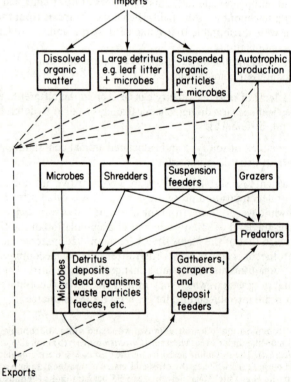

Fig. 7.4 Food resources and their utilization by freshwater decomposer communities. (From Berrie, A.D. (1976). pp. 323–38 in *The Role of Terrestrial and Aquatic Organisms in Decomposition Processes* (J.M. Anderson and A. Macfadyen, eds). Blackwell Scientific Publications, Oxford.)

Resource Quality

Resource quality is a composite definition of the value of food to an organism and embodies both physical and chemical criteria. The resource must satisfy the physical (surface properties, texture, etc.) and chemical (phagostimulatory, growth factors and nutrients) requirements for ingestion and colonization to occur. The chemical attributes of resource quality can be classified into three main groups of compounds: the carbon and energy sources, nutrients and modifiers.

Carbon and energy sources

The bulk of carbon and energy in dead organic matter is present as a variety of polymeric compounds of which the structural polysaccharides and lignin of plant remains form the major proportion. These compounds differ widely in their biodegradability. Most saprotrophs can catabolize simple sugars, starch and hemicelluloses but cellulose and lignin are more resistant.

Cellulose is an unbranched $\beta1-4$ linked polymer of glucose with a chain length in the region of $2-14 \times 10^3$ glucose units. The polymer chains are arranged in a crystalline lattice maintained by hydrogen bonds between the hydroxyl groups of neighbouring chains. The resistance of cellulose chains to enzyme attack is greatly increased by the crystalline organization, in rather the same way as a thick wad of paper is less readily burnt than the individual pages. The decomposition of cellulose involves the opening up of the crystal lattice as well as the depolymerization of the chains and is accomplished by a complex of several enzymes which are not widespread in either animal or microbial communities (see Swift *et al.*, 1979).

Lignin is a very complex and variable polymer, the structure of which is not understood. It is basically a polymer of coniferyl alcohol, p-coumaryl alcohol and sinapyl alcohol arranged in a variety of three-dimensional forms. The resistance of lignin to depolymerization is conferred not only by the stability of the phenyl rings and covalent bonding between side chains but also by its overall hydrophobic nature. These resistant properties of lignin can also result in an inhibition of enzyme attack on cellulose when the two polymers are in close proximity.

Minderman (1968) hypothesized that the pattern of weight losses from plant materials might be a summation of the decay curves for the individual constituents. Figure 7.5 shows loss rates (and percentage composition) for the main groups of compounds in woodland leaf litter plotted as logarithmic decay curves. Decomposition after one year ranged from almost 100 per cent losses of simple sugars to 50 per cent for lignin and as little as 10 per cent for phenols. The summation curve (S) does not conform to a simple logarithmic function because of the increased proportion of the

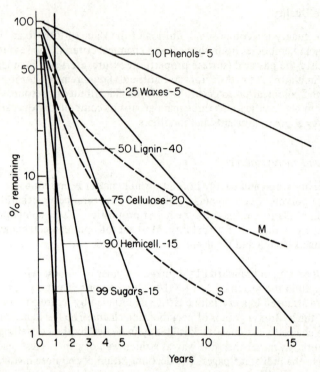

Fig. 7.5 The decomposition curves of the various groups of constituents, if their decomposition could be represented by a logarithmic function (the straight lines from the point 100%). The number in front of the name of the constituent indicates the loss after one year. The number after the constituent represents its percentage in weight of the original litter (these values are rough averages and they do not represent a specific analysis). The line S shows the summation curve obtained by annual summation of the residual values of the separate components. The line M gives an approximation, based on some analyses, of the probable course of the decomposition of similar resources in the mor-type forest soil at Hackfort. (Redrawn from Minderman, 1968.)

slower decomposing leaf components with time but the observed weight losses showed an even slower rate loss than that predicted. The most probable reasons for this are firstly that synergistic interactions occur between compounds and affect their decay characteristics, and secondly compounds synthesized by microorganisms may be more resistant than the primary leaf compounds. Nonetheless, this model provides an intuitive appreciation of the way in which the pattern of decomposition is determined by the constituent compounds.

Nutrients

The effects of limiting nutrient concentrations on saprotroph activity has been emphasized and focusses attention on nutrients as major components of resource quality. Table 7.3 lists macronutrient concentrations in a range of primary and secondary resource types. In general terms nutrient element concentrations are highest in animal and microbial tissues and then decrease through graminaceous and deciduous leaf litters, coniferous and ericaceous litters to the lowest nutrient concentrations in woody tissues. Decomposition rates would rank in approximately the same order but it is difficult to prove the causal basis of this relationship because of other correlated

Table 7.3 Nutrient element composition (% dry weight) of plant (primary) resources and decomposers (secondary resources). All leaf material after litter fall; root and wood from living plants. (From Swift *et al.*, 1979.)

Resource type	Species	N	P	K	Ca	Mg
Deciduous leaf	*Populus tremuloides*	0.56	0.15	0.60	2.35	—
	Betula papyrifera	0.58	0.32	0.78	1.71	—
Conifer leaf	*Abies casiocarpa*	0.69	0.09	0.30	1.18	—
	Pinus contorta	0.51	0.04	0.15	0.55	—
Grass leaf	*Eriophorum vaginatum*	0.97	0.04	0.09	0.20	0.08
	Nardus stricta	0.53	0.03	0.10	0.08	0.08
Shrub leaf	*Rubus chamaemorus*	1.31	0.07	0.09	0.85	0.53
Shrub shoot	*Calluna vulgaris*	1.38	0.07	0.09	0.34	0.06
Tree root large	*Quercus petraea*	0.5	0.06	0.2	0.4	0.08
Tree root small	*Q. petraea*	0.9	0.10	0.4	0.4	0.11
Grass root	*E. vaginatum*	0.50	0.06	0.21	0.11	0.08
Outer bark	*Q. petraea*	0.5	0.17	0.08	0.5	0.03
Cambium	*Q. petraea*	0.9	0.08	0.4	1.3	0.15
Sapwood	*Q. petraea*	0.16	0.02	0.14	0.05	0.01
Inner heartwood	*Q. petraea*	0.10	0.01	0.06	0.06	0.01
DECOMPOSERS						
Fungus mycelium (on leaf)	*Mycena galopus*	3.60	0.24	0.57	—	—
Fungus fruit bodies (on leaf)	Mixed	—	0.68	2.90	0.07	0.07
Fungi (on leaf)	Mixed	2.80	0.24	0.12	3.30	0.19
Fungus mycelium (on wood)	*Stereum hirsutum*	1.34	0.09	0.41	0.79	0.10
Fungus fruit bodies (on wood)		1.87	0.33	0.88	0.07	0.12
Bacteria – culture	Range	8–15	2–6	1–2	1	1
Bacteria – leaves	Mixed	4.0	0.91	1.50	0.95	0.15
Oligochaeta	—	10.5	1.1	0.5	0.3	0.2
Diplopoda	—	5.8	1.9	0.5	14.0	0.2
Insecta	—	8.5	6.9	0.7	0.3	0.2
Detritivores	—	7.74	0.80	0.13	10.30	0.27
Fungivores		7.74	1.39	0.40	3.95	0.46

resource quality effects. For example, Heath and Arnold (1966) showed that the thin soft leaves from shaded parts of trees, beech and oak in their study, decompose more rapidly than the heavily cuticularized leaves from the outer regions of the tree which are exposed to sunlight, wind and rain. King and Heath (1967) analysed the soft, medium and hard leaves and showed that the contents of ash, total nitrogen, tannins, sugars, lignin and cellulose all varied between leaf types. Any one of these parameters could therefore be significantly correlated with weight losses.

Carbon:nitrogen ratios (and other carbon:nutrient ratios) are frequently taken as a measure of biodegradability because they incorporate both the carbon and nutrient components of resource quality. Microbial tissues have a C:N ratio of about 10:1, but 25–30:1 is the optimum C:N ratio for a resource because some carbon is respired as the metabolic cost of tissue synthesis. When the C:N is higher than about 25:1 carbon is respired and lost from the organic matter while the nitrogen is converted to microbial protein. The C:N ratio decreases with time until it approaches 25:1 when the microbial tissues themselves are recycled. Under conditions of excess nitrogen (C:N lower than 25:1), nitrogen will probably be lost as ammonia. There is a broad correlation between C:N ratios and decomposition rates of organic materials but the relationship may be obscured by other limiting nutrients and the availability of nitrogen and carbon since both nutrients can be found in labile and inert pools.

The importance of C:N ratios has long been appreciated in agricultural practice. Figure 7.6 shows the availability to tomato plants of nitrogen added to the soil in organic form as plant debris over a four week period. Only with plant materials which had C:N ratios below about 20:1 was

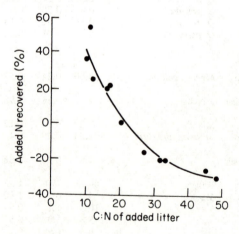

Fig. 7.6 Percentage recovery of nitrogen by tomato plants in four weeks from eleven different plant residues with a range of carbon:nitrogen ratios. (Redrawn from Black, C.A. (1968). *Soil Plant Relationships*, 2nd ed. Wiley, London.)

any nitrogen mineralized. At higher C:N ratios mineral nitrogen availability was decreased due to the immobilization of nitrogen, drawn from the soil solution, in microbial tissues.

Modifiers

Modifiers are chemical components of resources which influence the physiology or behaviour of decomposer organisms. The plant protection compounds, which have been evolved to reduce herbivore damage to plants, are still effective after leaf fall and the tannins in particular are implicated as important modifiers of decomposition processes. The same practical problems exist, as with the nutrients, of differentiating specific effects of tannins in multi-factorial situations but certain generalizations can be drawn (Swift *et al.*, 1979): (1) the primary mode of action is the formation of resistant complexes with proteins, consequently reducing nitrogen availability; (2) the complexes formed under acid, nutrient-poor conditions are more resistant than those produced under high pH and nutrient conditions; (3) they also act by direct inhibition of fungal and faunal activity; and (4) polyphenols are produced in the greatest concentration and diversity by plant species growing on nutrient poor soils, and may contribute to the formation of organic soils under these conditions.

Many pesticide and fungicide compounds used by man to replace unpalatable plant protection chemicals in agricultural crops also have modifying effects on decomposition processes (Edwards, 1970). Others are accidental pollutants, such as heavy metals from mining activities. Strojan (1978), for example, measured concentrations of about 26 000 ppm Zn, 10 000 ppm Fe, 2300 ppm Pb, 900 ppm Cd, 340 ppm Cu and 0.40 per cent S in the surface leaf litter layers of a woodland site one kilometre from a zinc smelter in Pennsylvalia (U.S.A.) A control site contained 650 ppm Zn, 2800 ppm Fe, 260 ppm Pb, 9 ppm Cd, 50 ppm Cu and 0.13 per cent S. The average standing crop of organic matter on the control site was 3.8 kg m^{-2} compared with about 8.1 kg m^{-2} in the site one kilometre from the smelter, suggesting a long-term depression of decomposition and mineral cycling. A similar study by Rühling and Tyler (1973), carried out under controlled laboratory conditions, confirmed the inhibitory effects of heavy metals on decomposition processes at even moderate concentrations. Copper, zinc, cadmium and nickel ions have strong fungicidal effects, particularly on spore germination.

Physical Rate Determinants

The effects of climate on decomposition processes, as well as physicochemical environmental factors such as aeration and pH, are expressed at scales relevant to the organisms concerned. The physicochemical determinants of decomposition processes at the microsite level are discussed by Swift *et al.* (1979) and involve detail which is outside the present account. The influence of temperature and moisture on decomposition will therefore be limited to consideration at the ecosystem level.

Olson (1963) suggested that a useful expression of the decomposition characteristics of ecosystems, where a long-term P:R equilibrium has been reached, is the index 'k' calculated as the ratio of total dead organic matter input (I) to the decomposer subsystem to the standing crop of organic matter (X_{ss}):

$$k = \frac{I}{X_{ss}}$$

It is rarely possible to calculate the overall ratio for ecosystems because living and dead roots are difficult to differentiate and contribute to both I and X. In addition, large woody litter input is variable in time and space. An alternative value of k is the ratio of above ground litter input (I_L) to the litter standing crop (X_L); k values can be similarly calculated for any litter component though precautions have to be taken over seasonal variations in standing crops. Values for k_L in major ecosystem types (Fig. 7.7) roughly parallel global patterns of terrestrial NPP and suggest the predominance of climate as a regulatory factor at this scale of organization. In humid rain forests temperature and moisture conditions are favourable for decompo-

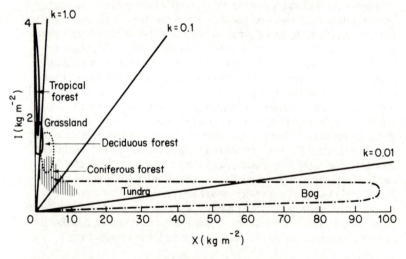

Fig. 7.7 Input (I) and accumulation (X) of organic matter in a number of biomes. The decomposition index k = I/X (for details see text). Inputs are best estimates of total input from primary production and accumulation is the total amount of organic matter from soil. (Redrawn from Heal, O.W. *et al.* (1981). In *Analysis of Ecosystems: Tundra Biome* (L.C. Bliss, O.W. Heal and J.J. Moore, eds.). Cambridge University Press.)

sition processes throughout the year and decomposition rates are generally faster than input rates. The k values for tropical forests are usually
greater than one and reflect rapid decomposition and generally low soil
organic matter. There are, however, exceptions to this generalization,
which is prevalent in the ecological literature, and several forest soil types
in the humid tropics show levels of organic matter accumulation similar
to, or even exceeding, those of temperate forests. In a kerangas (tropical
heath forest) soil in Borneo the standing crop of SOM (soil organic matter)
(270 t ha^{-1}) was comparable to the total forest biomass (340 t ha^{-1}), let
alone litter fall ($9.4 \text{ t ha}^{-1} \text{ yr}^{-1}$) and the k value for total organic matter in
this site would be similar to boreal forest or tundra (J.M. Anderson and
J. Proctor, unpublished data). Despite the apparently favourable physical
environment for decomposition, very low litter quality (low mineral
nutrients, high lignin and tannins) apparently limits the biomass and
activity of the decomposer community. Waterlogging and the inhibition
of decomposition by some clay minerals (particularly allophane) can also
lead to high levels of organic matter accumulation in tropical soils.

Temperate grasslands may have comparable k values to tropical forests
although climatic conditions are more variable. The resource quality of
graminaceous litter however, is higher than that of deciduous trees because
of lower lignin and polyphenol content. At the other extreme litter input
in the tundra is a fraction of tropical NPP but organic matter accumulation
is higher by several orders of magnitude. Temperature is the principal
factor limiting decomposition. Air temperatures rise above freezing for
only 80–90 days in the year, rarely exceeding $10°C$, and are below $-20°C$
for much of the winter. In summer the surface layers of the soil thaw to a
depth of about 40 cm below which there is permanent frost. Both temperature and low moisture conditions can limit decomposition on the soil
surface but low oxygen tensions, under waterlogged conditions, combine
with low temperatures to inhibit decomposition processes at greater depth
in the soil profile.

Clinal variation in soil organic matter in relation to temperature and
moisture gradients in the U.S.A. has been shown by Brady (1974). A
transect from Canada to the Southern States is associated with a marked
increase in mean temperature and organic matter in grassland soils of
similar type. An increase in annual rain fall from West to East, from the
prairie grasslands to the edge of the deciduous forests, also shows an increase in soil organic matter and may be associated with a gradient of
increasing NPP.

Meentemeyer (1974) showed a close correlation between annual evapotranspiration and the weight loss of leaf litter for ten different forest sites
in the U.S.A. (Fig. 7.8(a)). An equation derived from this relationship
shows a good predictive capacity for trends in weight losses for a number
of sites in a European deciduous forest (Netherlands) over a 6 year period
(Fig. 7.8(b)).

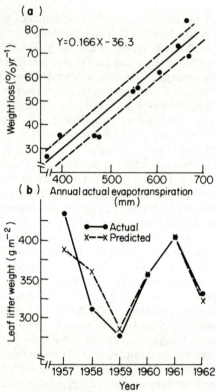

Fig. 7.8 The relationships between annual actual evapotranspiration and decomposition loss. **(a)** Plotted for first year weight loss and actual evapotranspiration for tree leaf litters in ten sites in the U.S.A. The regression equation is $Y = 0.166X - 36.3$. Standard errors of the regression line are indicated by the dotted line. **(b)** Yearly trends in the loss of weight by decomposition of leaf litter in a European deciduous forest (Netherlands) as observed (solid line) and predicted (broken line) by the above equation. (From Meentemeyer, 1974.)

Applied Aspects of Decomposition: Biological Waste Treatment

Organic wastes can be classified into two principal types — the dry solids (domestic refuse, wood waste, crop debris, etc.) and sewage. Until recently crop waste and animal excretion was recycled within the farm system as organic manures. Untreated human sewage, however, was disposed of in rivers as late as the 1880s in Europe and in estuaries until the late 1960s.

As a result of increasing livestock populations and rising standards of living the production of organic waste has increased proportionately faster than human populations and waste disposal has become a major problem. In most developed countries sewage is processed to reduce eutrophication, reduce health hazards and safeguard drinking water supplies. For refuse disposal the major methods have been various forms of tipping (land fill) and incineration. However, tipping sites are becoming limited and the capital and energy costs of incineration are high so that the future is likely to see a significant increase in the number of plants for composting refuse and dewatered sewage sludge (Gray and Biddlestone, 1974).

The following sections briefly outline the biological aspects of sewage and composting processes; details can be found in Hawkes (1963) and Warren (1971), and Gray and Biddlestone (1974) on which these accounts are based.

Sewage treatment

Sewage is a mixture of human and domestic stock excreta, products from food processing industries, industrial waste containing metal salts and organic polymers, and large volumes of drainage water. The composition of sewage is thus extremely variable not only from urban to rural areas but also according to rainfall. The solids, however, represent a high quality resource (Table 7.4) and therefore have high biodegradative potential under suitable conditions. Most modern sewage plants are therefore biological processes which utilize microorganisms to convert organic waste to gaseous products and humic residues. The biochemical reactions are summarized in Fig. 7.9 and are basically the same as those occurring in 'self-cleansing' of rivers, but localized in both space and time.

Table 7.4 Approximate analysis of sewage. (From Higgins and Burns, 1975.)

Lipids (ether soluble fraction)	30%
Amino acids, starch and pectins (H_2O soluble)	8%
Hemicellulose	3%
Cellulose	4%
Lignin	6%
Protein	25%
Alcohol soluble fraction	3%
Ash	21%

Sewage processing occurs as a series of sequential steps (Fig. 7.10) aimed at removing the solids and biologically oxidizing colloidal and dissolved materials. Trickling filters and activated sludge are the principal oxidation methods in modern sewage plants.

More than 80 per cent of the settleable solids are removed in the primary sedimentation tanks, including suspended organic matter flocculated by bacterial action, which would otherwise clog the trickling filters. The filters are the familiar circular beds of gravel or similar material on to which sewage is spread by a rotating arm. The beds are well drained to maintain aerobic conditions and the large surface area to effluent volume ensures a close contact between the organism community and the sewage as it trickles downwards through the filter.

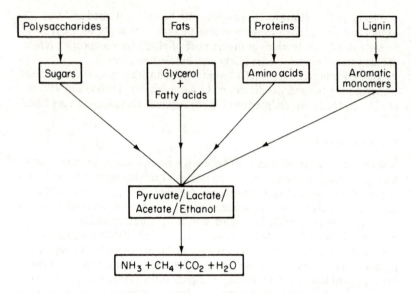

Fig. 7.9 The decomposition of organic matter during sewage treatment. (Redrawn, with permission, from Higgins and Burns (1975). Copyright by Academic Press Inc., London.)

The biological film in a trickling filter is a complex community of micro-organisms, primarily bacteria, protozoa, fly larvae worms and other invertebrates. The four common species of fungi have high production efficiencies, an undesirable feature where maximum respiration is required, and their growth can clog the filter beds (Hawkes, 1963). The animal populations normally regulate the microbial populations and reduce the occurrence of clogging. The community dynamics are complex and take many weeks to stabilize after a filter is set up. Some quantitative aspects of bacterial and protozoan interactions are explored by Curds (1973).

In the activated sludge process the primary sludge is pumped into aerated tanks and continuously innoculated with sludge (floc) from the secondary sedimentation tanks. The mixture is vigorously stirred to maintain the flocculated sludge in suspension and to ensure aeration. Pure oxygen may also be injected. The environment of an activated sludge system is biologically more specialized than a trickling filter and the diversity of microorganisms is lower. Few animal groups, including protozoa, can tolerate turbulent conditions. Those bacterial populations which are favoured by these conditions have near optimum resource requirements and have very rapid growth and biodegradation rates. Flocculent material in effluent from the filters and sludge tanks is allowed to settle in secondary sedimentation tanks. Various combinations of designs and sequences of treatment improve the effluent water quality (see Warren,

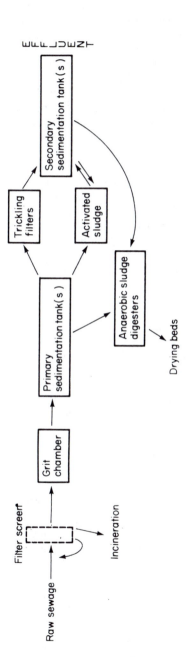

Fig. 7.10 Trickling filter and activated sludge treatment for sewage processing. See text for details. (Reproduced, with permission, from Higgins and Burns (1975). Copyright by Academic Press Inc., London.)

1971). Secondary treatments may be followed by a series of oxidation lagoons (tertiary treatments) which further reduce Biological Oxygen Demand, bacterial flocs and dissolved mineral nutrients, particularly nitrogen and phosphorus, through algal growth and incorporation into bottom sediments. These tertiary treatment ponds have been successfully employed for fish culture but the concentration of heavy metals from industrial effluents is a potential source of hazard.

The primary and secondary sludges can be dried and incinerated, composted or pumped into anaerobic digesters. The dried sludge can also be used as a fertilizer as most pathogenic bacteria are destroyed by the treatment but the presence of heavy metals in this material, particularly copper in pig slurry, has aroused concern in recent years (Higgins and Burns, 1975).

The anaerobic digestion process breaks down the bulk of the remaining organic matter into organic acids, methane and CO_2. The process takes about a month at $15°C$ and so disenters (digesters) are usually heated, using the methane they produce, to around $25-35°C$ which reduces the storage time to 20–30 days. Temperatures as high as $55°C$ are sometimes used where thermophilic* bacteria digest the sludge in 5–10 days but the high temperature process is difficult to control (Warren, 1971).

Composting

Municipal and commercial composting of domestic waste is not carried out extensively at the present time but is included here as it illustrates applied aspects of the variables influencing decomposition processes.

Some indication of the annual production of organic wastes in the U.K. is given by Table 7.5. These materials are extremely variable both in bulk and composition and a considerable improvement in the rate of composting can be achieved by optimizing nutrient conditions, particle size, homogeneity, moisture content, temperature and agitation (Gray *et al.*, 1971b). A process

Table 7.5 Approximate production of organic waste in the United Kingdom. (Data from Gray and Biddlestone, 1974.)

Source	Tonnes per year, fresh weight	Moisture content %, fresh weight basis
Wood-shavings and sawdust	1 070 000	—
Straws – wheat, barley, oats	3 000 000	14
Potato and pea haulms, sugar beet tops	1 600 000	77–83
Bracken, potentially available	1 000 000	—
Seaweed, potentially available	1 000 000	—
Garden wastes	1–10 000 000	—
Sewage sludge	20 000 000	95
Municipal refuse	18 000 000	20–40
Farm manures	120 000 000	85

*Thermophiles exhibit optimal growth above $40°C$ with upper thermal limits as high as $75°C$.

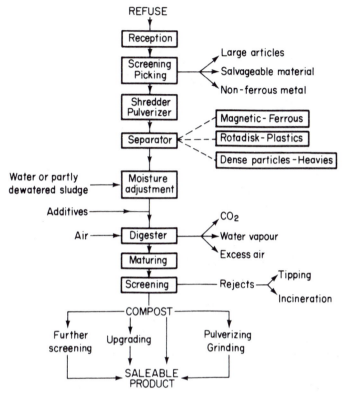

Fig. 7.11 Typical process flow diagram of a composting plant. (From Gray *et al.*, 1973.)

flow design of a complex municipal composting plant is shown in Fig. 7.11.

Particle size

Maceration of waste increases the surface of materials exposed to microbial attack and has a marked effect on the cellulolytic activity of thermophilic fungi over a temperature range of 45–60°C. The bacteria which
predominate at higher and lower temperatures, however, are utilizing
soluble carbon sources, such as sugars, and are less affected by the subdivision of the material. The optimum size range is approximately 1 cm
for forced aeration and agitated systems and 3 cm for static systems.
This size range is well above that which favours the bacterial growth form
but too small a particle size impedes the gaseous diffusion and could lead
to the formation of anaerobic conditions.

Nutrients

The relationship between C:N ratio of materials and the temperature

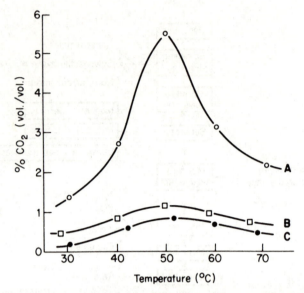

Fig. 7.12 Variation of decomposition rates (CO_2 production) with temperature for materials of various C:N ratios: **A**, grass containing abundant and readily accessible nitrogen, sugars and cellulose (C:N = 12); **B**, wheat straw containing large amounts of cellulose but little nitrogen (C:N = 128); **C**, sawdust which contains less cellulose and more lignin than straw but very little nitrogen (C:N = 500). (After Gray *et al.*, 1971b.)

response is shown in Fig. 7.12. Maximum decomposition rate is only achieved with adequate availability of nitrogen and carbon sources. In young grass (C:N = 19:1) the major proportion of the nitrogen is in available protein and the carbon as soluble sugars and cellulose. Microbial growth on sawdust (C:N = 500:1) is limited by low nitrogen and high lignin content and shows a reduced temperature response.

The optimum C:N ratio for composting is about 35:1 and although the ratio for combined refuse is close to this value it varies widely according to the proportional composition of the various types of organic waste (Table 7.6). Gray *et al.*, (1971a) consider that it is advantageous to optimize concentrations of both nitrogen and phosphorus in the feed to reduce residence time and maximize the quality of the final product.

The nutrient composition can be balanced by the addition of sewage sludge, suitable organic wastes, basic slag or fertilizers.

Moisture

Organic matter has a high water holding capacity because of its fine capillary structure. In the presence of excess moisture the spaces fill with

Table 7.6 Approximate nitrogen content and C:N ratios for some compostable materials (dry weight basis). (From Gray and Biddlestone, 1974.)

Material	N%	C:N
Urine	15–18	0.8
Blood	10–14	3
Mixed slaughterhouse wastes	7–10	2
Night soil	5.5–6.5	6–10
Activated sludge	5.0–6.0	6
Young grass clippings	4.0	12
Cabbage	3.6	12
Grass clippings (average–mixed)	2.4	19
Farmyard manure (average)	2.15	14
Seaweed	1.9	19
Potato haulms	1.5	25
Combined refuse	1.05	34
Oat straw	1.05	48
Wheat straw	0.3	128
Fresh sawdust	0.11	511
Newspaper	Nil	∞

water, oxygen diffusion is restricted and fungal growth is inhibited. This can also occur where aeration is restricted and metabolic water accumulates in the compost. When the moisture content in the feed is too low it can be raised by adding water or sewage sludge. The optimum moisture content is 50 per cent (net weight) for static heaps and 70 per cent for mechanically agitated systems with forced air injection where dessication can occur.

Time sequence of the composting process

The composting process can be divided into four stages – mesophilic, thermophilic, cooling and maturing.

At the start of decay the compost is at ambient temperature but as the mesophilic fungi and bacteria* grow the insulative properties of the organic material leads to the retention of metabolic heat and the temperature rises. Acid-producing bacteria are characteristic of this stage and the pH may drop slightly to around 6.0. As the temperature rises above 40°C the activity of the mesophiles declines and the thermophilic bacteria, and to a lesser extent the fungi, continue the reaction. The pH increases rapidly to over 8.0 and ammonia may be liberated if excess nitrogen is present. This first phase is associated with extensive cellulolysis, particularly by thermophilic fungi which reach peak activity at 55°C, but both cellulose and lignin are little attacked over 60°C when proteins, waxes and hemicelluloses are

*Mesophiles have optimum growth conditions between 25°C and 37°C and little growth occurs above 40°C.

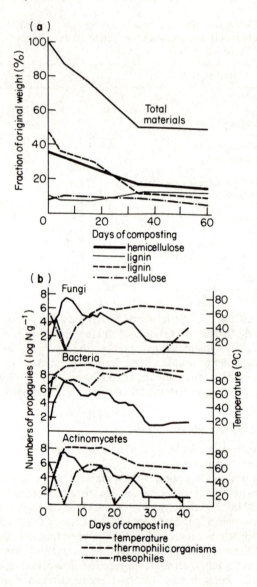

Fig. 7.13 Changes in the composition of **(a)** wheat straw and **(b)** microbial populations during the composting of wheat straw. ((a) From Yung Chang (1967). *Trans. Br. mycol. Soc.*, **50**, 667–77; (b) Yung Chang and Hudson, H. (1967). *Trans. Br. mycol. Soc.*, **50**, 649–66.)

readily degraded by thermophilic actinomycetes and other bacteria. Above 60°C the thermophilic fungi die off and the reaction is maintained by spore-forming bacteria and actinomycetes up to about 70°C. As the readily utilizable material becomes exhausted heat production declines and the compost cools. Once the temperature falls below 60°C (from a maximum of about 70°C) the thermophilic fungi can colonize the mass of compost from the cooler peripheral regions and continue cellulose decomposition.

The first three phases take from a few days to weeks but maturation, which results in the formation of stable humus complexes, can take several months.

Detailed analysis of microbial population changes or of biochemical changes have not been made during the composting of municipal waste but the above patterns are typical for decomposition processes in most bulked organic materials and are illustrated by the data for wheat straw composting shown in Fig. 7.13. Note that most cellulolysis occurs during the initial heating and cooling phases and not during the high temperature phase and that lignin is not only resistant to decay by this microbial community but increases proportionally as carbon mineralization progresses.

The products of composting processes have a wide variety of potential uses according to the quality. Gray and Biddlestone (1974) suggest that compost from farm or domestic wastes are valuable soil conditioners (provided that the problems of heavy metal contaminants can be overcome), mushroom bed material, high energy fuels for cement kilns and even valuable fibre for the preparation of building materials. There are also indications of whole new technologies developing around the recovery of energy and materials from waste – domestic gas and pyrolytic oils from municipal waste, fuels and chemicals from crop residues, oils and methanol from wood waste (Anderson and Tillman, 1977). A clear reaffirmation of the old saying that 'where there is muck, there is brass (money)'.

References

Important general texts, reviews and papers of particular interest are indicated by an asterisk.

*ALLABY, M. (1977). *World Food Resources: Actual and Potential.* Applied Science Publishers, Barking.

ANDERSON, J.M. (1973). The breakdown and decomposition of sweet chestnut (*Castanea sativa* Mill.) and beech (*Fagus sylvatica* L.) leaf litter in two deciduous woodland soils. I Breakdown, leaching and decomposition. *Oecologia* (Berl.), **12**, 251–74.

ANDERSON, J.M. (1978). Inter- and intra-habitat relationships between woodland *Cryptostigmata* species diversity and the diversity of soil and litter micro-habitats. *Oecologia*, **32**, 341–8.

*ANDERSON, J.M. and MACFADYEN, A. (eds) (1976). *The Role of Terrestrial and Aquatic Organisms in Decomposition Processes.* Blackwell Scientific Publications, Oxford.

ANDERSON, L.L. and TILLMAN, D.A. (eds) (1977). *Fuel from Wastes.* Academic Press, London and New York.

ARRHENIUS, E. (ed.) (1977). Nitrogen. *Ambio*, **6**, 95–182.

AUSMUS, B.S., EDWARDS, N.T. and WITKAMP, M. (1976). Microbial immobilization of carbon, nitrogen, phosphorus and potassium: implications for forest ecosystem processes. pp. 397–418 in *The Role of Terrestrial and Aquatic Organisms in Decomposition Processes.* J.M. ANDERSON and A. MACFADYEN (eds). Blackwell Scientific Publications, Oxford.

*BACASTOW, R. and KEELING, C.D. (1973). Atmospheric carbon dioxide and radiocarbon in the natural carbon cycle. pp. 86–136 in *Carbon and the Biosphere.* G.M. WOODWELL and E.V. PECAN (eds). A.E.C. Technical Information Centre, Washington D.C.

BAGENAL, T. (1978). *Methods for Assessing Fish Production in Freshwater*, 3rd edition. International Biological Programme Handbook 3. Blackwell Scientific Publications, Oxford.

BARKER, J.S.F. (1971). Ecological differences and competitive interaction between *Drosophila melanogaster* and *Drosophila simulans* in small laboratory populations. *Oecologia*, **8**, 139–56.

BELL, A.R. and NALEWAJA, J.D. (1968). Competitive effects of wild oat in flax. *Weed Sci.,* **16**, 501–4.

BERNAYS, E. and CHAPMAN, R. (1970). Experiments to determine the basis of food selection by *Chorthippus parallelus* (Zetterstedt) (Orthoptera Acrididae) in the field. *J. Anim. Ecol.,* **39**, 761–76.

*Biosphere (1970). Scientific American Books. Freeman, San Francisco.

*BOFFEY, P.M. (1971). Herbicides in Vietnam: AAAS study finds widespread devastation. *Science,* **171**, 43–7.

BONNER, J. (1962). The upper limit of crop yield. *Science,* **137**, 11–15.

BOWMAN, J.C. (1977). *Animals for Man.* Studies in Biology, no. 78. Edward Arnold, London.

*BRADSHAW, A.D. and CHADWICK, M.J. (1980). *The Restoration of Land.* Blackwell Scientific Publications, Oxford.

*BRADY, N.C. (1974). *The Nature and Properties of Soils,* 8th edition. Macmillan, New York.

*BRAEKKE, F.H. (ed.) (1976). *Impact of Acid Precipitation on Forest and Freshwater Ecosystems in Norway. Report of the SNSF Project.* Research Report FR6/76. SNSF, Oslo.

BRINKHURST, R.O. (1965). Observations on the recovery of a British river from gross organic pollution. *Hydrobiologia,* **25**, 9–51.

*BROWN, L.R. and ECKHOLM, E.P. (1975). Man, food and environment. pp. 67–94 in *Environment: Resources, Pollution and Society.* W. MURDOCK (ed.). Sinauer Associates, Stamford.

BURGES, A. and RAW, F. (eds) (1967). *Soil Biology.* Academic Press, London and New York.

*BUDYKO, M.I. (1974). *Climate and Life.* Academic Press, New York.

CARL, E.A. (1971). Population control in Arctic ground squirrels. *Ecology,* **52**, 395–413.

CAUGHLEY, G. (1970). Eruption of ungulate populations, with emphasis on the Himalayan thar in New Zealand. *Ecology,* **56**, 410–8.

CAUGHLEY, G. (1976). Plant-herbivore systems. pp. 94–113 in *Theoretical Ecology.* R.M. MAY (ed.). Blackwell Scientific Publications, Oxford.

*CHADWICK, M.J. and GOODMAN, G.T. (eds) (1975). *The Ecology of Resource Degradation and Renewal.* Blackwell Scientific Publications, Oxford.

CLATWORTHY, J.M. and HARPER, J.C. (1962). The comparative biology of closely related species living in the same area. V. Inter- and intra-specific interference within cultures of *Lemna* spp and *Salvinia natans. J. Exp. Bot.,* **13**, 307–24.

CLOUD, P. and GIBOR, A. (1970). The oxygen cycle. *Scientific American,* September, pp. 111–23.

COOKE, R.C. (1977). *Fungi, Man and His Environment.* Longman, London and New York.

COULSON, J.C. (1968). Difference in the quality of birds nesting in the centre and on the edges of a colony. *Nature,* **217**, 478–9.

COX, J.L. (1971). DDT residues in sea water and particulate matter in the Californian current system. *Fishery Bull.*, **69**, 443–50.

CREED, E.R., LEES, D.R. and DUCKET, J.G. (1973). Biological method of estimating smoke and SO_2 pollution. *Nature*, **244**, 278–80.

CROCKER, R.C. and MAJOR, J. (1955). Soil development in relation to vegetation and surface age at Glacier Bay, Alaska. *J. Ecol.*, **43**, 427–48.

CROMBIE, A.C. (1945). On competition between different species of graminivorous insects. *Proc. Royal Soc. Lond. Ser. B.*, **132**, 362–95.

CUMMINS, K.W. (1973). Trophic relations of aquatic insects. *An. Rev. Ent.*, **18**, 183–206.

CURDS, C.R. (1973). A theoretical study of factors influencing the microbial population dynamics of the activated-sludge process. 1. The effects of diurnal variations of sewage and carnivorous ciliated Protozoa. *Water Res.*, **7**, 1269–84.

*DALYRYMPLE, D. (1975). *Measuring the Green Revolution: The Impact of Research on Wheat and Rice Production.* USDA, Economic Research Service, Foreign Agricultural Economic Report No. 106 (July). Washington, D.C.

DECLAIRE, J.R. (1978). Economics of edible mushrooms. pp. 727–93 in *The Biology and Cultivation of Edible Mushrooms.* S.T. CHANG and W.A. HAYES (eds). Academic Press, London and New York.

DELWICHE, C.C. (1970). The Nitrogen Cycle. pp. 69–80 in *The Biosphere.* Scientific American Books. W.H. Freeman and Company, San Francisco.

DICKINSON, C.H. and PUGH, G.J.F. (eds) (1974). *Biology of Plant Litter Decomposition.* Vols 1 and 2. Academic Press, London and New York.

*DIXON, A.F.G. (1970). Quality and availability of food for a sycamore aphid population. pp. 271–88 in *Animal Populations in Relation to Their Food Resources.* A. WATSON (ed.). Blackwell Scientific Publications, Oxford.

DODD, A.P. (1940). *The Biological Campaign against Prickly Pear.* Government Printer, Brisbane.

*DUFFUS, J.H. (1980). *Environmental Toxicology.* Resource and Environmental Sciences Series. Edward Arnold, London.

EDMONDSON, W.T. (1969). Cultural eutrophication with special reference to Lake Washington. *Mitt. Int. Verein. Limnol.*, **17**, 19–32.

EDMONDSON, W.T. (1970). Phosphorus, nitrogen and algae in Lake Washington after diversion of sewage. *Science.* **169**, 690–91.

EDWARDS, C.A. (1970). *Persistent Pesticides in the Environment.* Chemical Rubber Co., Cleveland.

*EHRLICH, P.R., EHRLICH, A.H. and HOLDREN, P.R. (1977). *Ecoscience: Population, Resources, Environment.* W.H. Freeman and Company, San Francisco.

ELTON, C. (1927). *Animal Ecology.* Sidgwick and Jackson, London.

ESTES, J.A. and PALMISANO, J.F. (1974). Sea otters: their role in structuring nearshore communities. *Science,* **185**, 1058–60.

GADGIL, M. and SOLBRIG, O.T. (1972). The concept of r- and k-selection: evidence from wild flowers and some theoretical considerations. *Amer. Nat.,* **106**, 14–31.

GAUSE, G.F. (1932). Experimental studies on the struggle for existence. I. Mixed population of two species of yeast. *J. Exp. Biol.,* **9**, 389–402.

GILBERT, C.E. and RAVEN, P.H. (eds) (1975). *Co-evolution of Animals and Plants.* University of Texas Press, Austin.

GOLTERMAN, H.L. (1976). Zonation of mineralization in stratifying lakes. pp. 3–22 in *The Role of Terrestrial and Aquatic Organisms in Decomposition Processes.* J.M. ANDERSON and A. MACFADYEN (eds). Blackwell Scientific Publications, Oxford.

GORDON, A.G. and GORHAM, E. (1963). Ecological aspects of air pollution from an iron-scintering plant at Wawa, Ontario. *Canad. J. Bot.,* **41**, 1063–78.

GRANT, P.R. (1969). Experimental studies of competitive interaction in a two-species system. I. *Microtus* and *Clethrionomys* species in enclosures. *Can. J. Zool.,* **47**, 1059–82.

GRAY, K.R., SHERMAN, K. and BIDDLESTONE, A.J. (1971a). Review of Composting Part 1. *Process Biochem.,* **6**, 22–8.

GRAY, K.R., SHERMAN, K. and BIDDLESTONE, A.J. (1971b). Review of Composting Part 2. The Practical Process. *Process Biochem.,* **6**, 32–6.

GRAY, K.R. BIDDLESTONE, A.J. and CLARK, R. (1973). Review of Composting Part 3. Processes and Products. *Process Biochem.,* **8**, 7–12.

*GRAY, K.R. and BIDDLESTONE, A.J. (1974). Decomposition of urban waste. pp. 743–75 in *The Biology of Plant Litter Decomposition,* Vol. 2. C.H. DICKINSON and G.J.F. PUGH (eds). Academic Press, London and New York.

GRAY, T.R.G. and WILLIAMS, S.T. (1971). *Soil Micro-organisms.* University Reviews in Botany 2. Oliver and Boyd, Edinburgh.

GRIME, J.P., BLYTHE, G.M. and THORNTON, J.D. (1970). Food selection by the snail *Cepaea nemoralis* L. pp. 73–100 in *Animal Populations in Relation to Their Food Resources.* A. WATSON (ed.). Blackwell Scientific Publications, Oxford.

GRINNEL, J. (1917). Field test of theories concerning distributional control. *Amer. Nat.,* **51**, 115–28.

HAINSWORTH, F.R. and WOLFF, L.L. (1970). Regulation of oxygen consumption and body temperature during torpor of a hummingbird, *Eulampis jugularis. Science,* **168**, 368–9.

HARDING, L.W. and PHILLIPS, J.H. (1978). Polychlorinated biphenyls: transfer from microparticulates to marine phytoplankton and effects on photosynthesis. *Science,* **202**, 1189–92.

HARPER, J.L. (1969). The role of predation in vegetational diversity. *Brookhaven Symp. Biol.,* **22**, 48–62.

*HARPER, J.L. (1977). *Population Biology of Plants.* Academic Press, London and New York.

HARRISS, R.C., WHITE, D.B. and MACFARLANE, R.B. (1970). Mercury compounds reduce photosynthesis by plankton. *Science,* **170,** 736–7.

HASSELL, M.P. (1976). *The Dynamics of Competition and Predation.* Studies in Biology No. 72. Edward Arnold, London.

*HAWKES, H.A. (1963). *The Ecology of Waste Water Treatment.* Pergamon Press, Oxford.

HAWKES, H.A. (1975). River zonation and classification. pp. 312–74 in *River Ecology.* B.A.WHITTON (ed.). Blackwell Scientific Publications, Oxford.

HAYES, W.A. and LIM, W.C. (1980). Wheat and rice straw composts and mushroom production. pp. 85–94 in *Straw Decay and its Effect on Disposal and Utilization.* E. GROSSBARD (ed.). Wiley, Chichester and New York.

HEAL, O.W.H. and MACLEAN, S.F. (1975). Comparative productivity in ecosystems – secondary productivity. pp. 89–108 in *Unifying Concepts in Ecology.* W.H. van DOBBEN and R.H. LOWE-McCONNELL (eds). W. Junk, Hague.

HEATH, G.W. and ARNOLD, M.K. (1966). Studies in leaf litter breakdown. II. Breakdown rate of 'sun' and 'shade' leaves. *Pedobiologia,* **6,** 238–43.

*HIGGINS, I.J. and BURNS, R.G. (1975). *The Chemistry and Microbiology of Pollution.* Academic Press, London and New York.

*HOBSON, P.N. (1976). *The Microflora of the Rumen.* Meadowfield Press, Shildon.

HOLLING, C.S. (1959). Some characteristics of simple types of predation and parasitism. *Canadian Entomologist.* **91,** 385–98.

HOLLING, C.S. (1965). The functional response of predators to prey density and its role in mimicry and population regulation. *Mem. Ent. Soc. Can.,* **45,** 3–60.

HOLMES, W. (1977). Choosing between animals. *Phil. Trans. Roy. Soc. B.,* **281,** 121–37.

HOUSE, H.L. (1961). Insect Nutrition. *Ann. Rev. Ent.,* **6,** 13–20.

*HOUSE, H.L. (1966). The role of nutritional principles in biological control. *Can. Ent.,* **98,** 1121–34.

HUFFAKER, C.B. (1958). Experimental studies on predation: dispersion factors and predator–prey oscillations. *Hilgardia,* **27,** 343–83.

*HUGHES, R.D. and WALKER, J. (1970). The role of food in the population dynamics of the Australian Bush Fly. pp. 255–70 in *Animal Populations in Relation to Their Food Resources.* A. WATSON (ed.). Blackwell Scientific Publications. Oxford.

HUTCHINSON, G.E. (1957). *A Treatise on Limnology,* Vols I and II. Wiley, New York.

JAFFE, A. (1975). Ranching on the wildside. *International Wildlife,* **5,** 4–13.

JENSEN, N.F. (1978). Limits to growth in world food production. *Science,* **201**, 317–20.

KING, H.G.C. and HEATH, G.W. (1967). The chemical analysis of small samples of leaf material and the relationship between the disappearance and composition of leaves. *Pedobiologia,* **7**, 192–7.

*KREBS, C.J. (1978). *Ecology: The Experimental Analysis of Distribution and Abundance,* 2nd edition. Harper and Row, London and New York.

*KREBS, C.J. and MYERS, J.H. (1974). Population cycles in small mammals. pp. 267–399 in *Advances in Ecological Research.* A. MACFADYEN (ed.). Academic Press, London and New York.

LACK, D. (1944). Ecological aspects of species formation in Passerine birds. *Ibis,* **86**, 260–86.

LACK, D. (1966). *Population Studies of Birds.* Clarendon Press, Oxford.

LAWRENCE, D.B., SEHOUIHI, R.E., QUISPEL, A. and BOND, G. (1967). Role of *Dryas drummondii* in vegetational development. *J. Ecol.,* **55**, 793–813.

LEVIN, D.A. (1969). The role of trichomes in plant defence. *Quart. Rev. Biol.,* **48**, 3–15.

LEWONTIN, R.C. (1969). The meaning of stability. *Brookhaven Symp. Biol.,* **22**, 13–24.

LIETH, H. (1975). Global productivity in ecosystems: comparative analysis of global patterns. pp. 67–88 in *Unifying Concepts in Ecology.* W.H. van DOBBEN and R.H. LOWE-McCONNELL (eds). W. Junk, Hague.

LIETH, H. and BOX, E. (1972). Evapotranspiration and primary productivity; C.W. Thornthwaite Memorial Model. pp. 37–44 in *Papers on Selected Topics in Climatology.* J.R. MATHER (ed.). Elmer, New York.

*LIKENS, G.E., BORMANN, F.H., JOHNSON, N.E., FISHER, D.W. and PIERCE, R.S. (1970). Effects of forest cutting and herbicide treatment on nutrient budgets in the Hubbard Brook watershed ecosystem. *Ecological Monographs,* **40**, 23–47.

LIKENS, G.E., BORMAN, F.H., PIERCE, R.S. and REINERS, W.A. (1978). Recovery of a deforested ecosystem. *Science,* **199**, 492–6.

LINDEMAN, R.L. (1942). The trophic dynamic aspect of ecology. *Ecology,* **23**, 399–418.

*MACARTHUR, R.H. (1968). The theory of the niche. pp. 159–76 in *Population Biology and Evolution.* R.C. LEWONTIN (ed.). Syracuse University Press, Syracuse, New York.

MACARTHUR, R.H. and MACARTHUR, J.W. (1961). On bird species diversity. *Ecology,* **42**, 594–8.

MACHTA, L. and HUGHES, E. (1970). Atmospheric oxygen in 1967 to 1970. *Science,* **168**, 1582–4.

MARPLES, T.G. (1966). A radionuclide tracer study of arthropod food chains in a *Spartina* salt-marsh ecosystem. *Ecology,* **47**, 270–7.

*MAY, R.M. (ed.) (1976). *Theoretical Ecology: Principles and Applica-*

tions. Blackwell Scientific Publications, Oxford.

MEENTEMEYER, V. (1974). Climatic water budget approach to forest problems. II The prediction of regional differences in decomposition rate of organic debris. *Pub. Climatol.*, **27**, 35–74.

MELLANBY, K. (1980). *The Biology of Pollution*, second edition. Studies in Biology No. 38. Edward Arnold, London.

MENZEL, D.W. and RYTHER, J.H. (1961). Annual variations in primary production of the Sargasso Sea off Bermuda. *Deep Sea Res.*, **7**, 282–8.

MERREL, D.J. (1951). Interspecific competition between *Drosophila lunebris* and *Drosophila melanogaster*. *Amer. Nat.*, **85**, 159–69.

*MILES, J. (1979). *Vegetation Dynamics*. Outline Studies in Ecology. Chapman and Hall, London.

MILNER, C. and HUGHES, R.E. (1968). *Methods for the Measurement of the Primary Production of Grassland*. International Biological Programme Handbook 6. Blackwell Scientific Publications, Oxford.

MINDERMAN, G. (1968). Addition, decomposition and accumulation of organic matter in forests. *J. Ecol.*, **56**, 355–62.

MOSSER, J.C., FISHER, N.S., TENG, T.C. and WURSTER, C.F. (1972). Polychlorinated biphenyls: toxicity to certain phyto-plankters. *Science*, **175**, 191–2.

*MUELLER-DOMBOIS, D. and ELLENBERG, H. (1974). *Aims and Methods of Vegetation Ecology*. Wiley, New York.

*MURDOCH, W. (ed.) (1975). *Environment*. Sinauer, Sunderland, Massachusetts.

MURDOCH, W.W., EVANS, F.C. and PETERSON, C.H. (1972). Diversity and pattern in plants and insects. *Ecology*, **53**, 819–28.

NEWBOULD, P.J. (1967). *Methods for Estimating the Primary Production of Forests*. International Biological Programme Handbook 2. Blackwell Scientific Publications, Oxford.

NICHOLSON, A.J. (1957). The self-adjustment of populations to change. *Cold Spring Harbour Symp. Quant. Biol.*, **22**, 153–73.

*ODUM, E.P. (1969). The strategy of ecosystem development. *Science*, **164**, 262–70.

OLSON, J.S. (1963). Energy storage and the balance of producers and decomposers in ecological systems. *Ecology*, **44**, 322–31.

OVINGTON, J.D. (1965). *Woodlands*. English University Press, London.

PAINE, R.T. (1966). Food web complexity and species diversity. *Amer. Nat.*, **100**, 65–75.

PENMAN, H. (1970). The water cycle. pp. 37–46 in *The Biosphere*. Scientific American Books. Freeman, San Francisco.

PETRUSEWICZ, K. and MACFADYEN, A. (1970). *Productivity of Terrestrial Animals – Principles and Methods*. International Biological Programme Handbook 13. Blackwell Scientific Publications, Oxford.

*PHILLIPSON, J. (1966). *Ecological Energetics*. Studies in Biology No. 1. Edward Arnold, London.

PHILLIPSON, J. (1970). The 'best estimate' of respiratory metabolism and its applicability to field situations. *Pol. Arch. Hydrobiol,* 17, 31–41.

PHILLIPSON, J. (1973). The biological efficiency of protein production by grazing and other land-based systems. pp. 217–35 in *The Biological Efficiency of Protein Production.* J.G.W. JONES (ed.). Cambridge University Press, Cambridge.

PIANKA, E.R. (1967). On lizard species diversity: North American flatland deserts. *Ecology,* 48, 333–51.

PIELOU, E.C. (1975). *Ecological Diversity.* Wiley-Interscience, New York.

PIERROU, U. (1976). The global phosphorus cycle. pp. 75–88 in *Nitrogen, Phosphorus and Sulphur Global Cycles.* B.H. SVENSSON and R. SÖDERLUND (eds). SCOPE Report 7. *Ecological Bulletins,* Vol. 22. Stockholm: Swedish Natural Science Research Council.

*PIMENTEL, D. and PIMENTEL, M. (1979). *Food Energy and Society.* Resource and Environmental Sciences Series. Edward Arnold, London.

PIMENTEL, D., TERHUNE, E.C., DYSON-HUDSON, R., ROCHEREAU, S., SAMIS, R., SMITH, E.A., DENMAN, D., REIFSCHNEIDER, D. and SHEPARD, M. (1976). Land degradation: effects on food and energy resources. *Science,* 194, 149–55.

*POMEROY, L.R. (1970). The strategy of mineral cycling. *Ann. Rev. Ecol. Syst.,* 1, 171–90.

*POSTGATE, J.R. and HILL, S. (1979). Nitrogen fixation. pp. 191–213 in *Microbial Ecology. A Conceptual Approach.* J.M. LYNCH and N.J. POOLE (eds). Blackwell Scientific Publications, Oxford.

QUARLES, H.D., HANAWALT, R.B. and ODUM, H.E. (1974). Lead in small mammals, plants and soil at varying distances from a highway. *J. Appl. Ecol.,* 11, 937–49.

*RAISWELL, R., BRIMBLECOMBE, D., DENT, D. and LISS, P. (1980). *Environmental Chemistry.* Resource and Environmental Sciences Series. Edward Arnold, London.

REAY, P.J. (1979). *Aquaculture.* Studies in Biology, no. 106. Edward Arnold, London.

REINERS, W.A. and REINERS, N.M. (1970). Energy and nutrient dynamics of forest floors in three Minnesota forests. *J. Ecol.,* 58, 497–520.

ROSENSWEIG, M.L. and WINAKUR, J. (1966). Population ecology of desert communities: habitat and environmental complexity. *Ecology,* 50, 558–72.

ROYAMA, T. (1970). Factors governing the hunting behaviour and selection of food by the great tit (*Parus major*). *J. Anim. Ecol.,* 39, 619–68.

RÜHLING, A. and TYLER, G. (1973). Heavy metal pollution and decomposition of spruce needle litter. *Oikos,* 24, 402–16.

RYTHER, J.H. (1969). Photosynthesis and fish production in the sea. *Science,* 166, 72–6.

RYTHER, J.H. and DUNSTAN, W.M. (1971). Nitrogen, phosphorus

and eutrophication in the coastal marine environment. *Science,* **171**, 1008–13.

*SALISBURY, F.B. and ROSS, C.W. (1978). *Plant Physiology*, second edition. Wadsworth Publishing Company, Belmont, California.

SHAPLEY, D. (1973). Herbicides: AAAS study finds dioxin in Vietnamese fish. *Science,* **180**, 544–54.

SLOBODKIN, L.B. and SANDERS, H.L. (1969). On the contribution of environmental predictability to species diversity. *Brookhaven Symposia in Biology,* **22**, 82–95.

SMITH, M.W. (1969). Changes in environment and biota of a natural lake after fertilization. *J. Fish. Res. Bd Canada,* **26**, 3101–32.

*SÖDERLUND, R. and SVENSSON, B.H. (1976). The global nitrogen cycle. pp. 23–76 in *Nitrogen, Phosphorus and Sulphur Global Cycles.* B.H. SVENSSON and R. SÖDERLUND (eds). SCOPE Report 7. *Ecol. Bull.,* Stockholm.

SOUTHERN, H.N. (1970). The natural control of a population of Tawny Owls (*Strix aluco*). *J. Zool. Lond.,* **162**, 197–285.

SOUTHWOOD, T.R.E. (1976). Bionomic strategies and population parameters. pp. 26–48 in *Theoretical Ecology.* R.M. MAY (ed.). Blackwell Scientific Publications, Oxford.

*SOUTHWOOD, T.R.E. (1978). *Ecological Methods with Particular Reference to the Study of Insects*, second edition. Methuen, London.

SOUTHWOOD, T.R.E., BROWN, V.K. and READER, P.M. (1979). The relationships of plant and insect diversities in succession. *Biol. J. Linn. Soc.,* **12**, 327–48.

STENNER, R.D. and NICKLESS, G. (1974). Adsorption of cadmium, copper and zinc by dog whelks in the Bristol Channel. *Nature,* **247**, 198–9.

STEWART, B.A., PORTER, L.K. and JOHNSON, D.D. (1963). Immobilization and mineralization of nitrogen in several organic fractions of soil. *Soil Sci. Soc. Amer. Proc.,* **27**, 302–4.

STROJAN, C. (1978). Forest leaf litter decomposition in the vicinity of a zinc smelter. *Oecologia* (Berl.), **32**, 203–12.

*SWIFT, M.J., HEAL, O.W. and ANDERSON, J.M. (1979). *Decomposition in Terrestrial Ecosystems.* Blackwell Scientific Publications, Oxford.

THOMAS, A.S. (1963). Further changes in vegetation since the advent of myxamatosis. *J. Ecol.,* **51**, 151–83.

THORNLEY, J.H.M. (1970). Respiration, growth and maintenance in plants. *Nature,* **227**, 304–5.

TOWNSEND, C.R. (1980). *The Ecology of Streams and Rivers.* Studies in Biology No. 122. Edward Arnold, London.

ULLYETT, G.C. (1947). Mortality factors in populations of *Plutella maculipennis* Curtis (Tineidae: Lep.) and their relation to the problem of control. *Ent. Mem. Dep. Agric. Un. S. Afr.,* **2**, 77–202.

ULLYET, G.C. (1950). Competition for food and allied phenomena in sheep blowfly populations. *Phil. Trans. Roy. Soc. Lond. (B),* **234,** 77–174.

*UNDERWOOD, E.J. (1971). *Trace Elements in Human and Animal Nutrition,* third edition. Academic Press, London and New York.

*VARLEY, G.C., GRADWELL, G.R. and HASSELL, M.P. (1973). *Insect Population Biology.* Blackwell Scientific Publications, Oxford.

VOLLENWEIDER, R.A. (ed.) (1969). *A Manual on Methods for Measuring Primary Productivity in Aquatic Environments.* International Biology Programme Handbook 12. Blackwell Scientific Publications, Oxford.

*WALTER, H. (1973). *Vegetation of the Earth in Relation to Climate and Ecophysiological Conditions.* English Universities Press, London.

WANGERSKY, P.J. and CUNNINGHAM, W.J. (1956). On time lags in equations of growth. *Proc. Nat. Acad. Sci. U.S.,* **42,** 699–702.

WARREN, C.E. (1971). *Biology and Water Pollution Control.* Saunders, London.

WARREN, V.H. (1972). Biogeochemistry in Canada. *Endeavour,* **31,** 46–8.

WATERS, T.F. and CRAWFORD, G.W. (1973). Annual production of a stream mayfly population: comparison of methods. *Limnology and Oceanography,* **18,** 286–93.

WATSON, A. and JENKINS, D. (1968). Experiments on population control by territorial behaviour in red grouse. *J. Anim. Ecol.,* **37,** 595–614.

WETZEL, R.G., RICH, P.H., MILLER, M.C. and ALLEN, H.L. (1972). Metabolism of dissolved and particulate detrital carbon. *Mem. Ist. Hal. Idrobiol.,* **29,** Suppl., 185–243.

*WHITTAKER, R.H. (1975). *Communities and Ecosystems,* second edition. Macmillan, New York.

*WHITTAKER, R.H., BORMANN, F.H., LIKENS, G.E. and SICCAMA, T.C. (1974). The Hubbard Brook ecosystem study: forest biomass and production. *Ecological Monographs,* **44,** 233–54.

*WHITTAKER, R.H. and FEENY, P.P. (1971). Allelochemics: chemical interactions between species. *Science,* **171,** 757–70.

WHITTAKER, R.H. and WOODWELL, G.M. (1969). Structure, production and diversity of the oak-pine forest at Brookhaven, New York. *J. Ecol.,* **57,** 157–76.

*WHITTON, B.A. (ed.) (1975). *River Ecology.* Blackwell Scientific Publications, Oxford.

WIHLM, J.F. (1975). Biological indices of pollution. pp. 375–402 in *River Ecology.* B.A. WHITTON (ed.). Blackwell Scientific Publications, Oxford.

WILLIAMS. R.J.H. and RICKS, G.R. (1975). Effects of combinations of atmospheric pollutants upon vegetation. pp. 127–38 in *The Ecology of Resource Degradation and Renewal.* M.J. CHADWICK and G.T. GOODMAN (eds). Blackwell Scientific Publications, Oxford.

WOLFF, T. (1977). Diversity and faunal composition of the deep-sea benthos. *Nature,* **267**, 780–5.

WOODWELL, G.M. (1970). The energy cycle of the biosphere. pp. 25–36 in *The Biosphere.* Scientific American Books, Freeman, San Francisco.

WOODWELL, G.M., WURSTER, C.F. and ISAACSON, P.A. (1967). DDT residues in an East Coast estuary: a case of biological concentration of a persistent pesticide. *Science,* **156**, 821–4.

WURSTER, C.F. (1968). DDT reduces photosynthesis by marine phytoplankton. *Science,* **156**, 821–4.

ZADRAZIL, F. (1980). Screening of Basidiomycetes for optimal utilization of straw. (Production of fruiting bodies and feed.) pp. 139–46 in *Straw Decay and its Effect on Disposal and Utilization.* E. GROSSBARD (ed.). Wiley, Chichester and New York.

*ZARET, T.M. and PAINE, R.T. (1973). Species introduction in a tropical lake. *Science,* **182**, 449–55.

Index